They Were Ready: The 164th Infantry In the Pacific War, 1942-1945

Terry L. Shoptaugh

Valley City, ND: 164th Infantry Association
of the United States

Published in the United States
All Rights Reserved

ISBN 978-0-615-35045-5

Printed In Canada

For information on the 164th Infantry Regiment:
Editor, 164th Infantry News
PO Box 1111, Bismarck ND, 58502-1111
TheyWereReady@hotmail.com

Published by the 164th Infantry Association

TABLE OF CONTENTS

Acknowledgements		iv
Maps		ix
1	Welcome to the Green Hell	1
2	Hard Times	28
3	In the Army Now	52
4	New Caledonia, "A Lazy Man's Paradise"	76
5	"These Farm Boys Can Fight"	97
6	"The Din Was Terrific"	124
7	Feeding the Mamolo	150
8	"Malaria, Exhaustion, Malnutrition"	180
9	Bitter Sweet Victory	208
10	A Brief Respite	231
11	Perimeter Defense on Bougainville	254
12	Victory by Attrition	278
13	"The Ones Who Slap You Back Are Filipinos"	305
14	Into the Breach Again, and Again	332
15	"The Greatest War in History"	358
16	Comrades	381

Acknowledgements

The war in the Pacific "was something no one not here can ever know," wrote a captain of the U.S. Marines in 1944, just after the end of the battle for the island of Saipan. Eleven days after writing this the captain was killed. But his words echoed the feelings of thousands of Americans who fought in the Pacific. It was a war unlike any other in the nation's experience, requiring enormous forces to traverse the almost limitless ocean and fight dozens of battles just to come into striking distance of the opponent's home country. It was a no-quarter war that required ordinary young men to kill even wounded foes who refused to surrender. When the men who fought this war came home, they believed no one but those who took part in it could really understand what it was like. Most of them said little about it to anyone who was not there. As one man said to his daughter when she asked about the war, a civilian "didn't need to hear about those things."

Sixty years after the war their feelings had changed. The passage of time and the attentions given to World War II, had loosened the resolve of the veterans to keep silent. If they did not speak now it would all be lost. So many did share their memories, some in written memoirs, others in interviews.

In 2006 I attended a dinner at the State Historical Society of North Dakota and met Lt. Col. Shirley J. Olgeirson (ret.), the editor of the newsletter for the 164th Infantry Association veterans group. She told me about the many veterans who could describe their experiences in the Pacific. "Many of them have great and valuable stories," she assured me, "and it has to be done soon."

She was right – the interviews began in the winter of 2007 and the stories were valuable, for the 164th regiment had fought across the Pacific from Guadalcanal to the Philippines. They had fought in war at its worst and had demonstrated the valor of the American soldier at his very best. By mid-2008, with over fifty in-person and telephone interviews, supplemented by tapes and transcripts of interviews conducted by others, letters and diaries of veterans who had passed on, the regiment's official records, and hundreds of photographs provided by the veterans. My aim in telling this story is to use the men's words as possible, providing enough context to explain what they had contributed to the victory. I hope that I have done that in a way that amply underscores their sacrifices and triumphs.

I owe a host of debts to those who helped me along the way. Most important are of course all of those men and women who granted me interviews, patiently answered my numerous inquiries and sent me information and photographs. I have tried to acknowledge these debts in the notes. Shirley Olgeirson helped me locate veterans for interviews and uncovered many of the letters and diaries that proved so valuable. James Fenelon, former editor of the *164th Infantry News*, kindly gave me copies of interviews he did with more than a dozen men during the 1990s, and shared with me some of his own memories of serving in the regiment's headquarters company. Colonel David Taylor (ret.), the National Commander of the Americal Division Association, shared valuable information with me and commented on portions of the manuscript. Robert Dodd Jr. gave me a copy of his book describing his father's service in the war and shared copies of the interviews he recorded while writing it. Wes Anderson of the Barnes County Historical Society was invaluable to me, in filming the interviews I conducted during the 2007 reunion of the regiment, and also in giving me copies the museum had of interviews with other veterans. Both the State Historical Society of North Dakota and the staff of the Libby Archives at the University of North Dakota were most generous in allowing me long term access to interviews and papers from their collections. Mary Fran Timboe Riggs was generous with items from her father's papers, while Todd Morgan sent me a copy of the tribute to Rilie Morgan. Warren Freeman, Scott Legaard and Leatrice Clauson Cooper also provided very helpful information about some of the men. The book could not in fact have been written without such kindnesses.

Michael Lyons, emeritus professor of history and the author of two fine books on the world wars, read early drafts of my chapters and gave me generous advice. Colonel Richard Stevens (ret.) not only shared with me his memories of service on Guadalcanal but also gave me the benefit of his knowledge in explaining the complicated American command structure during that long campaign. Several of the veterans, including John Paulson and Rudolph Edwardson, saved me from making errors in parts of the battle narratives; any remaining mistakes in the text are my sole responsibility.

As always I could not have completed this manuscript without the help of my assistant Korella Selzler, who proofed the chapters and helped me organize the illustrations. My wife, Deborah Janzen, once again supported me throughout the entire process. Finally, I have to acknowledge the inspiration I received from James M. Denney, my oldest friend, who years ago sparked my interest in the history of the war era. So too, in a curious way, did my father, Don C. Shoptaugh, who did his war service with the 26th Infantry, 1st Infantry Division, in France and Germany. Like so many others he said very little about it when I was growing up. Because of this, finding and preserving the stories of other "silent warriors" became one of my ambitions. I like to think of this work as one of the results.

Terry L. Shoptaugh

Maps

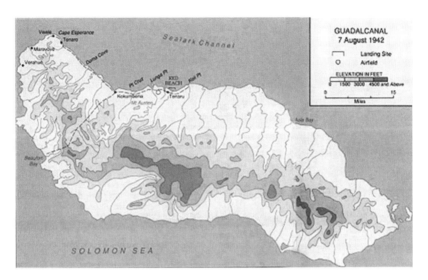

Map 1 – Guadalcanal, where the U.S. Marines landed in August 1942. The Japanese base at Rabaul, northwest of Guadalcanal, could reinforce their garrison on the island and bomb the Marines. Guadalcanal was barely within the reach of New Caledonia by American multi-engine aircraft in use at that time (DC-3s, B-17s, B-24s, and PBY naval flying boats).

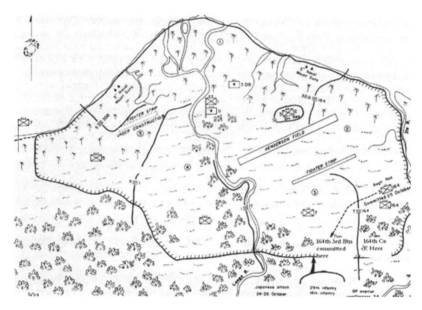

Map 2 -- A US Marine Corps sketch map of American positions on Guadalcanal in October 1942. The principal defensive perimeter was anchored on the Ilu River in the east (which the U.S. thought at the time was the Tenaru River). Entrenched positions ran from there south and then west, across the Lunga River and back north to the beach. Marine outposts further west, out to the Matanikau River, were in place to keep Japanese artillery further away from Henderson airfield and also to give early warning of an attack from the west. The 164th positions were centered on the southeast corner of the perimeter (the "Coffin Corner") and then west to meet the 7th Marine positions east of the Lunga. It was in this area that Japanese infantry launched heavy attacks from October 24 to 27. The position occupied by John Stannard's rifle platoon and Charles Walker's machine gun crew was right at the boundary line between the 164th Infantry's sector and the sector held by the 7th Marines, as marked on the map, lower right.

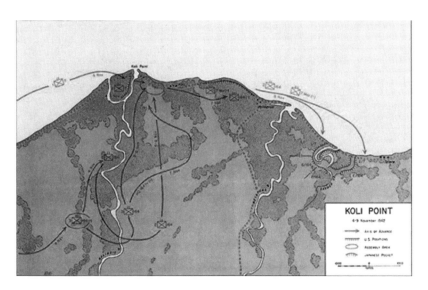

Map 3 -- The march to Koli Point and the battle to trap the Japanese around Gavaga Creek (From Miller, *Guadalcanal*).

Map 4 -- Sketch map of Koli Point, which renders the terrain and march routes to Koli (164th Infantry Association Records).

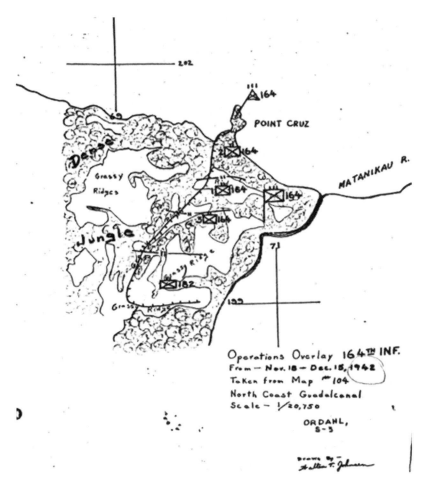

Map 5 – Sketch map of positions of the 164th on the Matanikau Front, from the regimental records, 164th Infantry. Within the area marked "grassy ridges" were the various hills where the men took heavy losses while attacking enemy positions.

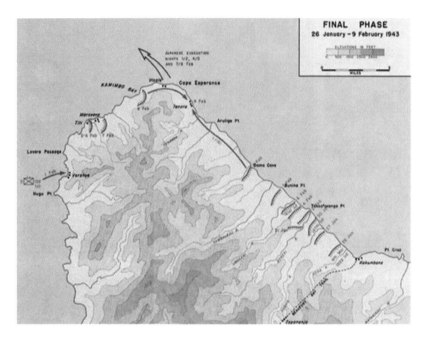

Map 6 – The last phase on Guadalcanal and the evacuation of Japanese forces (from Miller, *Guadalcanal*).

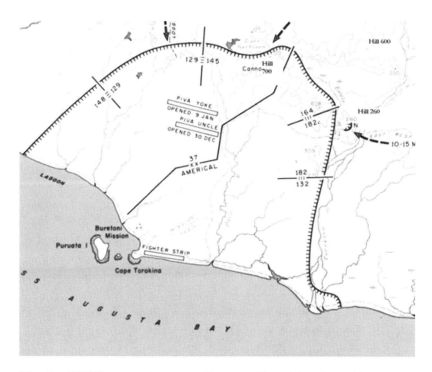

Map 7 -- XIV Corps perimeter on Bougainville, with unit positions marked as of March 1944, when the Japanese army launched its counteroffensive in three waves. Note the position of the 164th Infantry on the right flank of the 37th Infantry Division, northern tip of the perimeter. Hills 260, 600 and 700, all sites of major battles, are marked on the map. (Map source: U.S. Army, published in Miller, *Cartwheel: The Reduction of Rabaul*.)

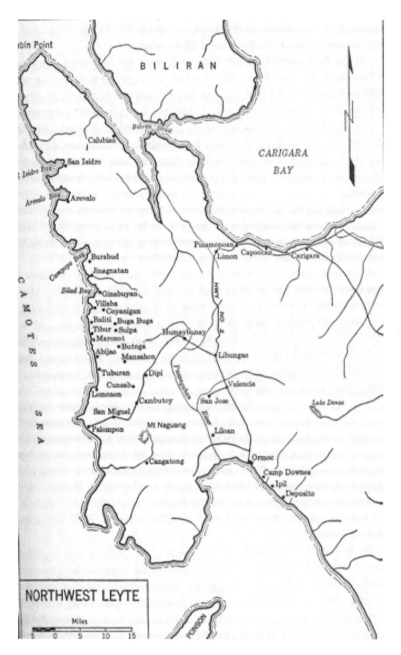

Map 8 -- The 164th's first battlefield in the Philippines, in northwest Leyte. Valencia is just north of the port of Ormoc (map from Cronin, *Under the Southern Cross*, p. 227).

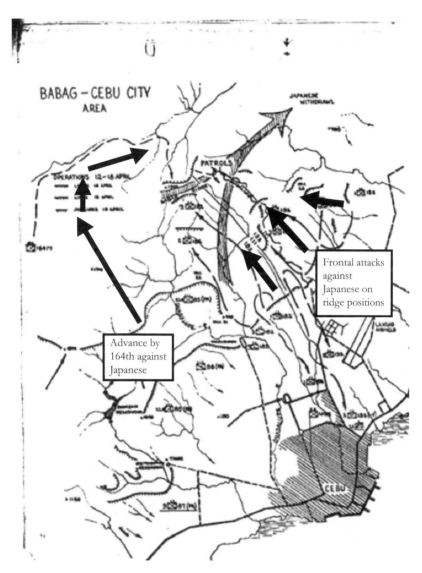

Map 9 -- As the 132nd and 182nd pinned General Manjome's Japanese forces in the hills north of Cebu City, the 164th marches north and east to encircle the enemy. (Map from "After Action Report, Americal Division, V2 Operation," 164th Infantry Association Records.)

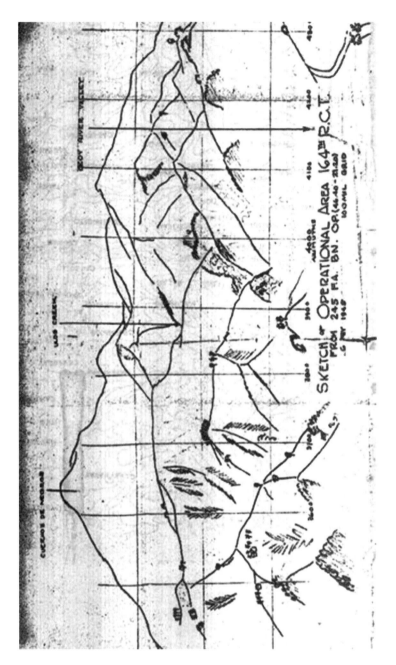

Map 10 -- Sketch made by artillery observers, showing Mount Cuernos de Negros and surrounding hills, on which the 1st and 2nd Battalion of the 164th waged a six week battle against entrenched Japanese units. Trees suggest expanse of the battlefield. (Sketch from S3 "Operations Report, Cebu/Negros.")

Chapter 1: Welcome to the Green Hell Guadalcanal, October 1942

From the decks of the *McCawley*, Guadalcanal looked like a moviegoer's idea of a south Pacific paradise. The waters of Sealark Channel were a rich blue; the waves sparkled in the sunlight. Sunrise spotlighted the green hills leading up to Mount Austen. The palm trees in the Lever Brothers plantation, stirred by the breeze, waved a welcome to the soldiers of the 164th Infantry Regiment.[1]

But the sailors on the deck of the navy transport *McCawley*, and those on its companion the *Zeilin*, knew that the clear skies were perfect for an attack by Japanese aircraft out of Rabaul. They wanted to land the soldiers quickly and be gone. To expedite the unloading, each platoon in the regiment detailed a man to work on the ship. "What you unload is what you get," the *McCawley's* captain told Sergeant Herschel Lawson of E Company, adding "unload quickly," for he planned to withdraw if any bombers threatened his ship. The first men to land now began to board the Higgins boats. Harry Wiens, a rifleman with Company I, debarked from the *Zeilin*. "We went down the ship's ladder, a stair-like device lowered alongside the vessel," Wiens wrote later. "As we debarked the ship, we heard the sounds of a few random artillery rounds being fired at some distance." Most of the 164th did not have the luxury of using stairs to board their landing craft. They climbed down cargo nets hung along the sides of the *Zeilin* and *McCawley*. Some of the sailors near the rails gave the "dogfaces" quick instructions on how to climb down the nets and drop into the boats, a maneuver some of the men had never even seen before.

They were weighted down with equipment, including the new-style helmets, which had just been issued, and the new Garand M1 rifle. With these and with their combat packs, gas masks and extra ammunition, more than a few of the men feared what would happen if they mistimed their jump for the landing craft. "It's a long drop to the bottom of the channel," a swabbie said, smiling.[2]

Soon, the first wave of men was on its way to the beach. Coming in, some soldiers wondered if they were going to have to face a Japanese attack the moment they went ashore. But, as Wiens put it, there was "nothing really to disquiet the serenity of the sunrise. Our boat took us to a spot just to the left of Lunga Point where it drove its flat bottom right to the beach, and as the ramp dropped, the water was scarcely ankle deep." The coxswains landed the first wave, then returned to the transports with two units of marines that were being withdrawn with a small handful of Japanese army prisoners, almost all of them Korean labor troops.

The men of the 164th landed on Guadalcanal at a critical moment. The First Marine Division had been there since August 7. Surprising the small contingent of Japanese on the island that morning, they had driven them into the jungle and quickly moved forward to seize the unfinished airfield they were building. The airfield had triggered the American attack; if completed, its planes could sever the link between the Americans and Australia and allow the Japanese to invade the French colony of New Caledonia. Thus the U.S. Navy landed the marines on Guadalcanal to take the airfield and develop it as an American base for a counterattack. The marines had since repelled every attack launched by the Japanese in their efforts to recapture that same airfield. But by October, the marines were badly worn down. Casualties, jungle illnesses and lack of supplies had reduced their number of effective fighters while the enemy, holding control of the waters around Guadalcanal with its larger navy forces, kept reinforcing its forces. The Japanese were now poised to begin another offensive.

The American Navy, much reduced by the losses at Pearl Harbor, had so far proven itself unable to keep the island properly supplied. American military leaders were expressing doubts as to whether Guadalcanal could be held. Men in the 164th heard rumors about some of these doubts back on New Caledonia. It was said that General Douglas MacArthur in Australia wanted Guadalcanal abandoned and its American troops transferred to his command. According to other rumors, Admiral Robert Ghormley, the commander of the South Pacific Area, the man ultimately in charge of the Guadalcanal campaign, had "written the campaign off" as an inevitable failure. Even Admiral Nimitz, the commander of the entire Pacific Ocean Area, told reporters that his navy seemed "unable to control the sea in the Guadalcanal area." Asked if the island could be held, Nimitz answered that the situation was "not hopeless, but it certainly is critical." The First Marine Division commander, Major General Alexander Archer Vandegrift, made it clear to Colonel Bryant Moore, the 164th commander, that he would fight to the last man, no matter what the "higher ups" believed. Vandegrift expected the GIs to do so as well.[3]

All of the Army reinforcements were ashore by about 7:30 a.m. Some of the GIs were detailed to begin organizing the incoming supplies and equipment, aided by a few Seabees and some of the off duty marines. As they hauled and stacked the heavy crates, regaining their land-legs and cursing the difficulty of moving over the soft sand, other 164th soldiers began to look around. Wiens and others from his platoon "walked inland a few hundred yards within the palms, and grinned on passing a small shack printed in large letters, 'TOJO'S ICE PLANT, UNDER NEW MANAGEMENT.'" They could also see that past the beach, the island didn't look quite so alluring. Unfilled shell holes, trees shattered by Japanese artillery fire, wrecked aircraft on the edges of the airfield, and other debris gave clear indication that this was not a set in some Hollywood production but a deadly serious combat zone. The First Marine Division cemetery was quite sobering.

Moving inland, the newcomers began to notice that the air was stifling. Mosquitoes swarmed upon the newcomers. Previous bombings and artillery shelling by the enemy had reduced trees and foliage to flinders. Henderson Field, the airfield, was the focal point of all the fighting. Wrecks of American aircraft were sitting under camouflage nets about the edge of the field. Hopefully Japanese raids would mistake these for operational planes and bomb them again. Tactically-minded officers may have noticed that the airfield was surrounded by higher ground. Enemy observers on the hills to the south and west could see just about everything inside the American perimeter.

A smell of decay permeated the entire area. "You could smell the stink of the rotting vegetation everywhere," Douglas Burtell commented. As part of the regiment's headquarters section, Burtell went immediately to the

Marine headquarters. "I was pretty good with maps, both making them and reading them, and knowing how to travel with them, and stuff like that. So I was sent to the First Marine intelligence section to get the dope on the maps." As he walked on, he saw broken and useless equipment along the trails. Small groups of marines came up to welcome the soldiers. "There was a nervous, excited exuberance [among the marines]," Harry Wiens commented. "They were tremendously glad to see us and eager to tell their stories of [their victories at] Tenaru and Bloody Ridge, and we were avid listeners." Some of the marines gave the soldiers small souvenirs that they had collected on the island. They also spoke about the enemy that the newcomers would soon be meeting on the field of battle. "For the first time we learned, firsthand from participants, that the Japs were not the super-warriors depicted in British memos," Wiens noted. "That they could be and had been beaten, most decisively in some instances." But, the marines warned, "the Geneva Convention was not part of our enemy soldier's training, as Japan had never signed it . . . we were facing a deadly, treacherous, implacable foe . . . a ruthless, merciless foe, totally indoctrinated to unswerving obedience."[4]

The 164th soldiers couldn't help but notice how fatigued many of the marines looked, and how young. "A lot of them were just kids, sixteen or seventeen," Alvin Tollefsrud recalled. Tollefsrud, a tall and spare platoon sergeant in Company L, who would himself be all of twenty-four in two more months, also concluded that their tough talk wasn't an act. "They were scrappers, that was for sure." For their part, the marines were impressed by the soldiers. A sergeant with the 11th Marines artillery commented that the newcomers "carried everything with them in what we called sea bags – they called them duffel bags. We couldn't believe all the stuff they were carrying. They were big, raw-boned boys too, from the Dakotas. They made us look puny." The sergeant was particularly interested in the M1 rifles most of the GIs carried. He expected that it would help swing the battle in the Americans' favor.[5]

The marines assured Harry Wiens and his buddies that the "Japanese tactics were lousy – like charging a machine gun with bayonets while following their sword-waving officers ... all this in a formation so tight it almost resembled close-order drill." Bill Hoffman, a member of Wiens' platoon, was talking to a small group of marines when he noticed a peculiar looking "looped wire attached to his belt upon which were strung circular objects that resembled dried apricots." "These are Jap ears," the young Marine told him. Other marines showed Wiens tobacco sacks filled with "Jap gold teeth ... this brought home vividly the primal, primitive level upon which the war was being waged." The message was clear: the 164th boys must themselves show no mercy.[6]

Back at the beach, Alvin Tollefsrud was helping to sort out the boatloads of supplies. "We got busy and unloaded Higgins boats and loaded trucks and got the stuff inland." Russell Schmoker, another rifleman in Company L, also helped to unload. "There was a Marine colonel down there in charge," he remembered. "All of a sudden the Japanese artillery dropped a shell and it knocked two marines down. We all hit the dirt. The colonel yelled 'come on, get up, they can't hit right here again!'" The boats moved about 75 yards down the beach and the Jap gun couldn't reach that far." The unloading continued until around 11:30 when a wave of Japanese bombers was detected by radar. As some fifty American fighters roared off Henderson Field to meet the incoming strike, the soldiers decided it was time to look for some kind of shelter.

Bennie Thornberg, a member of Company F, recorded in his diary that by his watch the bombing started at "11:30 sharp: that was the first bomb that I ever heard come down and it is one that I never will forget." As he ran to find shelter, Thornberg noted that moment as "the starting of the war for the 164th." Harry Wiens and a friend jumped into "a shallow depression in the sand" together with two marines who had used it during previous air attacks. "It was large enough to accommodate us four and we sat there talking until one of the marines said 'There they are,' and pointed to a neatly compact pattern of anti-aircraft bursts over the ocean in the direction of Tulagi Island." Most of the newly arrived soldiers had no idea where any of the properly prepared air raid shelters were located. They found cover in ditches along trails or in any low ground they could find.[7]

Tollefsrud, Schmoker, and others on the beach found cover in low ground. Schmoker got a laugh from watching one of the sailors he had met on the voyage up from New Caledonia. "He was a little guy from Mandan [North Dakota]. He wanted to see some action, he said. Well, about that time [in the bombing] a couple of Zeroes came over and strafed the beach and that was the last time I saw him. He ran back to his ship, went below deck and never came out again."[8]

Chuck Walker was a newly promoted First Lieutenant in charge of a weapons platoon in Company H. He and several of his men were unloading cases of gelignite explosives at the beach when the bombing began. Anxious to put some distance between themselves and the explosives, they moved toward a group of hardwood trees where they found "scant cover" from the bombs dropped by the Japanese twin-engine "Betty" bombers. Almost all of the bombs fell on or close to Henderson Field. The bombers had ignored the ships out in the channel completely, but the GIs still helping out on *McCawley* and *Zeilin* worried that the navy might still suddenly take off "south for friendlier waters" with them still on board.[9]

On the beach everyone went back to work and squad leaders began counting heads. At first it looked like the regiment had gotten off with only a few light wounds, but then word spread that one man had been killed by a bomb landing in the coconut trees southeast of Henderson. Medics who rushed to the grove of trees found the body up against one of the tree trunks and identified the man as Kenneth Foubert, a platoon motor corporal. He was killed by bomb fragments. Taking Ken's body to the Marine aid station manned by Navy corpsmen, the medics removed his dog tags and prepared the body for burial. The clerks at regimental headquarters duly recorded Foubert's name as the "first member of the regiment killed on Guadalcanal."[10]

Word of Foubert's death went through the ranks fairly quickly. It was accompanied by remarks that Kenny "hadn't kept his head down, he wanted to look at the airplanes." Perhaps this was true; certainly others, even as the marines yelled to them to "stay down," had not resisted the temptation to look up and watch the planes coming over. It was a comfort for the still-green GIs of the 164th to feel that Foubert had made a mistake and paid for it with his life. Albert Wiest, of Company M, didn't want "Ken's parents to hear this story." Wiest himself had taken cover in a low spot during the bombing but there was little protection and he knew he could have been killed just as easily.

Ken Foubert's death had an immediate impact. Typical for a guard outfit, the 164th was built around companies that knew one another well, often for years prior to the outbreak of the war. Bill Tucker, a machine gunner in Company M, went to high school with Foubert in Grand Forks, North Dakota. "I went to school with Kenny and most of those fellows in the company. I think Kenny played on the football team with me, but I'm not sure if he graduated." Born in 1921 in Westhope, North Dakota, Ken was one of several children of Israel and Helen Foubert. The family had moved to Grand Forks when Ken was very young. There, Israel Foubert found work installing and repairing plaster lathe in homes and businesses. The family had to carefully manage every penny, and Ken started work when still quite young. It's likely he dropped out of school to get a job. He did a number of odd jobs to make extra money for his family, and in March 1939, like many of his fellow guardsmen, was lured by the dollar per week he would get for joining.

Photographs taken of Ken in his Guard uniform show a pleasant faced man with slightly protruding ears, a diffident smile and a hint of shyness. Together with his comrades, Ken left North Dakota to enter Federal service in February 1941. When Pearl Harbor was attacked and war came to America, Ken knew that his "year of service" was now a commitment to fight until the war ended. A lot of the men grumbled about this but accepted the situation for what it was, an emergency in which they would put their lives on the line. They did it because "that was the job we had to do."[11]

Ken had traveled 9000 miles to fight for his country, only to die a few hours after setting foot on Guadalcanal. A second Japanese air raid struck the island around 2:00 p.m., wounding some more men. The target again was Henderson Field. Watching the bombing from a hillside, Harry Wiens "clearly saw craters erupt on the runway" as the Betty bombers passed over. Japanese howitzers, heavy guns that had just been brought to the island and entrenched west of the Matanikau River, opened up shortly afterwards, again with the aim to keep Henderson Field out of action. As the shelling went on, Ken Foubert was buried by a small detail in the cemetery the marines had laid out near Kukum Point. One of the many things that the transports had brought up from New Caledonia with the regiment was a shipment of white wooden crosses to use in marking the graves. Prior to this the marines had made wooden markers as needed. Now the first of the new wooden crosses was used for Ken. Just a few days later, a War Department telegram informed Foubert's family of his death.[12]

In his diary, Lieutenant Colonel Sam Baglien, the regiment's executive officer, recorded a second death at day's end: "at 6:00 p.m. the area was shelled by enemy artillery located west along the beach toward Point Cruz. The troops sought cover and, although alarmed, maintained order. Pvt. Park E. Jagears, Company "D" was killed. At 11:00 p.m. the regiment began movement toward bivouac areas about two miles east, between Lunga Point, Teneru River and Henderson Air Field. What a day and what a reception for our first day!"[13]

But the day's events were not yet over, not even close.

The 164th came to Guadalcanal with only a handful of men who had experienced combat. One of these was Robert Hall, the commander of the 3rd battalion. A tall, spare man who seldom smiled, Hall had seen war at its worst in the trenches of France in 1918. Albert Wiest remembered how Hall had told them at Guard exercises that nothing they had read or heard or imagined could prepare them for the reality of combat. "He liked to say to us 'you dudes don't know what's it's really like.'" Hall used the term "dude" as an old westerner used it, to identify someone who was still an amateur. "He'd say, 'you dudes will know more after you've seen the real thing.'" As evening came on that first day on Guadalcanal, the "dudes" were about to see the real thing.

John Paulson was part of one Company F rifle squad. He and many others expected that the Japanese navy would make an appearance their first night on Guadalcanal. "We had been told [in briefings] that our ships would pull out and the Jap navy would sail up and down the coast at night." Marines he met when he arrived told him and others, with some bitterness, that "our navy cuts and runs" before dark. So when Paulson saw the *McCawley* and *Zeilin* turn and make steam southward, he wondered how long it would be before they had visitors.[14]

Admiral Isoroku Yamamoto, the commander of his nation's fleet and architect of the Japanese attack on Pearl Harbor, had promised to mount an all-out air and sea offensive to wipe out the America perimeter on Guadalcanal. The American positions ran about six miles in length from east to west across the north shore of the island. It was a little more than two miles deep. Substantial Japanese ground reinforcements were now on the island. They waited for Yamamoto's ships and planes to knock out Henderson Field before they struck at the American lines.

The two air strikes on October 13 were only a small part of this offensive plan. A much heavier blow, in the form of a task force of two Japanese battleships accompanied by destroyers and a light cruiser, was already on its way. The *Kongo* and the *Haruna*, the two battleships, were on schedule to close on Guadalcanal shortly after midnight, October 14. Armed with hundreds of fourteen inch shells designed for bombarding land targets, the big ships were going to pour hellfire onto the Americans. A second Japanese task force containing fast transports was following the bombardment force, to land another 4500 infantry and artillery reinforcements on the island. These men, supported by air strikes and naval bombardments, were ordered to overwhelm the American forces on Guadalcanal.

The Japanese ships entered Sealark Channel almost precisely on schedule. Japanese soldiers on the island guided the big ships in by setting oil fires that pointed toward the American positions. The battleships also sent a float plane aloft to help direct the bombardment. At around 1:00 a.m., the *Kongo* cut loose with incendiary shells in order to mark the primary target: the airfield. As soon as the *Kongo's* captain was satisfied that he had Henderson in his crosshairs, he ordered both battlewagons to commence firing with armor-piercing and high explosive ammunition. Fourteen inch munitions, supplemented by smaller rounds from the Japanese destroyers, began to rain down on the Americans.

John Hagen was a communications specialist with the 3rd battalion headquarters. He had spent the day down by the beach in order to learn how to link his army lines into the marine communications. "I'd had to stay down at the beach because some idiot assigned me to the shore party switchboard. [As supplies came onto the beach] I had to call companies involved to tell them their stuff was here. They would race down to the beach with a truck and pick it up to take it to their assigned areas." Hagen was still at the beach late that night when he was visited by Major Harry Tenborg, the 3rd battalion's executive officer. "Tenborg and I were sitting on the side of this ring of sacks around the [fox] hole. A marine was now running the switchboard and I was watching him. When I took over I was going to disconnect all of his stuff and plug in our stuff. His was World War I vintage equipment and ours was 'World War I and half' stuff." Waiting to do this, Hagen looked out toward the channel. "I saw these three red flashing lights moving toward us. They looked like airplane lights except they were red. I said 'Major, those three lights coming this way can't be airplanes . . .' The marine took one look and said 'Jesus Christ' and jumped out of that hole and took off for the woods. Those 'red lights' were 14-inch shells coming in from a big Jap battlewagon . . . They hit right in front of us in the water. You talk about getting wet and sand raining all over you.'"[15]

Doug Campbell was a member of the 164th Regimental band and like everyone else he was trying to adjust to being at Guadalcanal. Fortunately there was an old veteran nearby to get him through the next few hours. "We spent the whole day unloading ships, getting all our stuff on shore. After that we were sitting around under coconut trees and so forth. In my particular case, I was sitting alongside our C.O. and warrant officer, Gerald Wright. I looked up the beach and there was an American flag up there and I thought, boy, that's reassuring to see an American flag, because we knew that they were present, other troops. About that time a shell lit on the beach and kicked up a lot of sand. And I said to this warrant officer, the C.O., I said, 'Look at that. There's a shell hit the beach up there.' He said, 'Well, don't worry about that because it's our Navy out there, and they're going to have a lifting barrage. They're going to work up into the hills where the Japs are.' I thought, that's fine, but about a minute after that, another shell lit about halfway between where the first one lit and where we were sitting. Well, I says, 'they're going the wrong way with their barrage.' He looked very puzzled. After that the shells started to come right into the area where we were at and we started taking over. In my particular case, I started looking for a hole and I saw one and I made a dive for it, but another fellow got there first and I landed on top of him."[16]

That was when Campbell realized that "I'd forgotten my rifle. I left it back under the coconut tree where I was sitting with the warrant officer. I figured well, I've got to get in a hole, but I've got to get my rifle first. So I crawled back, because everything was coming in then. I got my rifle and I headed for a hole. I found a dugout with a bunch of marines in it. And I got down in there and the marine says, 'Say, you got an ammo belt on you?' I said, 'Yes.' He said, 'Get it out of here.' He said, 'The concussion will set it off and we'll all get killed.' So I did that and he says. 'You got your rifle loaded?' and I says, `Yes.' He says, 'Well, take the clip out, because we don't want that in there. The concussion can set that off.' Then he says, 'You got your helmet strap buckled?' And I said, 'Yes.' He says, 'Take it off, you're liable to get your neck broke from the concussion.' Well, that was on-the-job training, and damn fast. Well, we stayed in the hole while everything went on all night. I guess they gave that to us for about three hours, and in the morning, when we crawled out of these holes, that whole area was devastated from the shells."

Others found themselves in even nastier predicaments. Chuck Walker was lying next to a shallow slit trench trying to get to sleep when the first salvo landed. Rolling over the ground to get into the trench, he found it already occupied, so he headed off to find other cover next to a Japanese truck that had been captured and put to use. But the next salvo came down near the truck, "shaking it violently." So Walker and several others rushed to the site of a Japanese latrine, now covered with lumber. Tearing the lumber off, they dove in to get below the ground; better to stink and live than be blown apart.[17]

PFC Bill Welander was standing in a grove of palm trees outside the 3rd battalion headquarters tent. He had recently been transferred from his original posting with Company G to the 3rd battalion headquarters because "I'd had a year of ROTC and was in my second year of college [when the Guard was federalized]. So they had put me in the headquarters as part of intelligence." Welander and a buddy were heading toward the headquarters tent just as the Japanese bombardment began. "We were just about a block-length away from the other [HQ] guys, and at first we thought it was our guys shooting toward the sea, but it was the other way around. They were dropping them right over us." It was too dangerous to try to make it to the shelter near the HQ, so Welander and his friend found some cover among the trees and hugged the ground as coconuts and palm fronds crashed down.[18]

A great many other men were forced to take cover in the open. Albert Wiest had taken his men in Company M over and past the airfield to bed down for the night. "We thought we were done for the night. Nearby, Jap laborers had built some narrow shelters, about 35 or 40 feet long, log covered, with little benches on each side [of the shelter]. That took care of quite a few guys," who had stretched out on the benches and lay on the floors of the shelters. When the bombardment began, the rest of the men, including Wiest, headed for those shelters. "There was no more room in any of these, except to crouch down as close to the entrance as you could. Then if you could hear one coming in too close, you could dive in on top of somebody and maybe get out of the way." A few shells hit close by and shook the shelter, dirt falling from the sandbags. A medic who had squeezed up against the shelter entrance told Wiest: "I know where there's a bunker that's empty and it's not very far from here, if you want to make a break for it." Wiest figured "anything was better than sitting here, half in the open," so he and two others followed the medic, crouching down while the man looked for the shelter in the darkness. "Well, he got lost! He said, 'dammit, that oughta be right around in here,' looking through the palm fronds. He said 'wait for the next shell burst so I can get a flash.' Well, we had to wait while about half a dozen of those monster shells came screaming overhead! He finally spotted it, and there wasn't a soul in it except for a two-foot-long iguana. We chased him out, and that's where I spent the rest of that night."

Doug Burtell had his own long night. When finished his chore to obtain maps, he was attached temporarily to a marine recon unit. There he met Clifford Fox, who would become a good friend. "He was in the First Marines [Division]." Burtell spent most of the remainder of the day with Fox, and was talking with him that night. Fox, Burtell noted, "had a little one-man slit trench. Well, first a plane came over and dropped some flares to mark the targets, and then they started firing star shells. Anyhow, we both got into this slit trench that he had, and all hell broke loose. I laid on top of him and it was a terrible night. Some wounded marine who was out of his head came running by us and we pulled him down on top of us. He had been hit in the face by gravel or shrapnel; the next morning I had blood all over the back of my helmet. There were pieces of shrapnel as big as you can imagine flying all over the place. There was one kid not very far from us, when it got daylight we saw his gas mask, a big chunk of shrapnel had driven it right into the ground until just the straps were still showing. If that had hit him it'd have cut him right in half."

John Paulson was standing in line to get a meal when the battleships opened fire. "We were over by the Ilu River, had our mess kits out to get something to eat. All of a sudden, 'boom, boom, boom.' The river had a bank about so high [three feet] and I jumped over it into the river, to at least get below ground level. After a while, I got to thinking – 'wonder if there's any alligators in this swampy, smelly thing?' I found out later that by golly there were [crocodiles]. So I crawled out of there and found some guys who had room in a dugout. So I got in there. It was hot, packed full, stuffy and sweaty. After a while it got quiet and a couple of us got out to get air. We didn't know they were just turning around and coming back. We were sitting out there, nice and cool, and all of a sudden, 'shahooom,' a big old coconut tree [about a hundred feet away] disappeared just like that. It was pretty nice back in that hole after seeing that."

Rudy Edwardson, a rifleman and supply corporal, also with F Company, saw another man elude death by a hair. "We'd probably got a quarter mile off the beach by nighttime and were probably lucky that's as far as we got because most of the shells went over us. We had one land in our area. A guy was over there, he was a big bruiser of a guy and was lucky he was laying with his legs spread out. It landed kind of between his legs and threw him up in the air and knocked him cuckoo for a while. But it didn't kill him. It was armor piercing evidently and didn't explode."[19]

Over with I Company, Harry Wiens lay flat on the "bare earth" and tried to present as small a target as possible. He managed to get his entrenching shovel off his belt and then made an effort to "scrap[e] a bit of earth" away under him, but ultimately he gave it up as pointless because "the barrage continued so unrelentingly intense that mostly you could only cling to earth and attempt to melt into it." During the lull in the shelling, as the Japanese ships reversed their courses, the I Company commander, Ralph Knott, pulled together several men and ordered them to the beach to meet a possible enemy landing. The men got there just in time for the next round of shells to come in. They had to hit the ground again and bury their faces in the sand. No one was hurt, but years later, one of them told Harry Wiens that he never did completely forgive Knott for sending them down to the open beach.

Men sitting in the reinforced shelters were just as frightened. Over at the regimental headquarters shelter, Sam Baglien was talking to "Col. Brookes of the New Zealand Army [who] stated this shelling was worse than [his experience on] Crete." One of the unit's chaplains, who was also in the shelter with Baglien, "straightened up in the corner of the hole and looked us all over and said, 'Gentlemen, I've done all *I* can for you.' This broke the tension; we giggled." Ed Mulligan, another member of the regiment's headquarters section, sat in another shelter praying until at one point he decided the Japanese would just keep firing until "all of us were killed." Then he just prayed for "a quick death."[20]

The officers at the marine divisional headquarters, much closer to Henderson, felt the impact not only of the shells but of exploding fuel and ammo depots ignited by the explosions. Sitting in his shelter, the division's commander, General Vandegrift, watched the walls shaking with each new blast. He wondered if the whole thing might cave in on him. His operations officer turned to him and said "I don't know how you feel but I think I prefer a good bombing or artillery shelling." Vandegrift tried to reply, saying "I think I do ...", but his words were cut off as a near miss knocked everyone to the ground. Vandegrift later wrote that someone who had not experienced such a mauling "cannot easily grasp a sensation compounded of frustration, helplessness, fear and, in the case of close hits, shock." The general was angered by his inability to fight back. "We owned no night fighters; our artillery could not reach the ships . . . Japanese ships insolently claimed the sea."[21]

The bombardment lasted until 3:00 a.m. Then the Japanese, having expended over a thousand rounds of ammunition, turned back for Rabaul. Most of the shaken Americans waited until the sun came up before they ventured out. Many were bleeding from the ears, a result of concussions. Vandegrift left his shelter earlier than most others. He went directly to the airfield to inspect the damage. "It was a terrible mess. The Pagoda [a name given to a Japanese-made building that was used for the air headquarters] was half wrecked and Roy [Geiger, commander of the island air units] told me he was going to knock it down since he figured the enemy was using it for a registration point. The shelling cost us forty-one killed." Virtually all the aircraft were out of action. Several dozen marines had been killed and wounded.

Leaving the airfield, Vandegrift went down to the beach. The shells had severed scores of telephone lines, and he visited with several men, including John Hagen, to discuss the urgency of restoring communications to his front lined units. Men spent most of the day cleaning up, trying to get a few airplanes back into service, and taking cover while Japanese planes attacked again and again. Everyone knew that the Japanese task force loaded with reinforcements was on its way. These transports entered the channel early in the morning of October 15, out of range of the American guns. Vandegrift and his staff were beside themselves with anger to see "enemy transports calmly discharging troops and supply off Tassafaronga [Point] to the west." A few patched-up American planes got airborne and damaged a couple of the Japanese ships, but all the enemy reinforcements got ashore and disappeared into the jungle. Japanese airplanes arrived to bomb the airfield again. Enemy artillery again lobbed shells into the American perimeter. Where was the American navy? The frustrated Vandegrift fired off a radio message to New Caledonia, angrily demanding "maximum support of air and surface units," without which the island would be lost.

Among the soldiers in the 164th there was mixture of emotions at the welcome they received. They were angry toward the navy, which they believed had done nothing more than drop them off and "bug out." They were uncertain about the coming days. Some felt despair and a sense of helplessness. There was fear. A man couldn't sit through the kind of ear-splitting, teeth-rattling pounding that the battleships delivered without feeling some of that. Two more men in the regiment had died as a result of the bombardment. Bernard Starkenberg, a long-time guardsman and warrant officer and one of the most liked men, died after a shell tore off his legs. Corporal Rollie Andrick from the 2nd battalion headquarters was killed by another shell. Several wounded men and a few suffering from shock were evacuated and flown back to New

Caledonia. Sergeant Frank Doe, a former member of the Massachusetts National Guard who had been transferred to the 164th to beef up its regimental headquarters staff, had been slightly wounded in the shelling. But as he waited to be treated by the medics, he looked over at the "really badly wounded men." He felt lucky and even a little guilty to be bothering the overwhelmed medical staff.[22]

In all, it had been a horrifying experience, even for the hardened veterans. Robert Leckie, a marine rifleman who would write one of the best first-hand accounts of this critical campaign, admitted that, after ten weeks of combat, this was "the night I nearly panicked." As the Japanese shells fell and exploded, he woke from a "deep sleep," and tried to get out of his foxhole. "I clawed frantically at my mosquito net. I tried to butt my way through it, tried to bull through gossamer." After a few minutes he calmed down and shook off "the panting pretzel into which panic had twisted me." Herbert Merillat, a young lieutenant in the First Marine Division's press unit, thought this pounding was the worst shock of his life. "When it was over, I felt as if I should seek out a newly arrived North Dakotan and tell him, 'Look, it's really not always this bad.'"[23]

Before long however, most of the men in the regiment bounced back from the "night of the battleships." John Hagen, remembering the rumors that the "navy had written off Guadalcanal," talked with some of his pals about their prospects. They all decided that, whatever happened, they "wouldn't surrender like at Bataan." It would be better to die than become a prisoner of the Japanese, they agreed. Hagen went to work to repair phone lines and link the battalion communications into the marine's network. Platoon lieutenants gathered their men and put them to work, and the marines took GI riflemen up to the lines so they could familiarize themselves with the terrain. Marine cooks fired up their cook stoves, using some of the newly arrived supplies to make pancakes. They invited the soldiers to help themselves. Seabees went to work patching holes in the airfield's tarmac. This was very dangerous work, especially if the holes were made by artillery fire, as one Seabee explained. "A Jap five-inch gun lobs a shell over on your airstrip and blasts a helluva hole. What are you going to do? You know, just as that Jap artilleryman knows, that if he leaves his gun in the same position and fires another shell, the second shell will hit in almost the same spot as the first one." The Seabees did their jobs anyway, knowing full well the dangers. "All you can do is depend on hearing that second shell coming and hope you can scramble far enough away before it explodes. But this is a gamble which is frowned upon by life insurance companies."[24]

Japanese bombers came again to hit the airfield. They also dropped some propaganda leaflets which urged the Americans to admit that their navy had abandoned them and use the attached "ticket to armistice" to give up. Like the marines before them, the GIs began using these as toilet paper. By evening, companies of the 1st and 2nd battalion were taking up positions at the perimeter line, replacing marine units dug in along the Tenaru River, and on the ground south of the airfield. The 3rd battalion was placed in reserve, its companies to be sent in as needed once a Japanese attack unfolded. Everyone was going about their business, and Baglien was moved to write "the troops had [had] no time to dig in [before the shelling], but sought all available cover and maintained good order" throughout the ordeal. He was confident that they would do a good job in the days ahead. Robert Leckie thought the "dogfaces" he encountered seemed naively confident. "They came up [to the perimeter] after another air raid; a very close one. But the Thing had not infected them yet. War was still a lark. Their faces were still heavy with flesh, their ribs padded, their eyes innocent. They were older than we, an average twenty-five to our average twenty. I remember when two of them, having heard of the Ilu [River], immediately set off for it, picking their way through the barbed wire, like botanists off on a field trip."[25]

But how would they act once they came face to face with the enemy, Leckie and his buddies wondered. Did they understand what would be required of them? When the Japanese launched one of their patented night attacks on the lines, come screaming out of the dark, would the GIs stand and fight or run? How tough were they? The answers to most of these questions would not be answered on October 14. All that the men of the 164th knew for certain after one day on the island was that they were in it for real now, and they were a long way from home.

(Endnotes)

[1] Background sources for this chapter are: Francis D. Cronin *Under the Southern Cross: The Saga of the American Division* (Washington: Combat Forces Press, 1951), pp. 50-51; Richard B. Frank, *Guadalcanal* (New York: Penguin Books, 1992), pp. 313-321.

[2] Harry Wiens, "My Little Corner of the War" (typed manuscript memoir, copy in 164th Infantry Association Records, Orin G. Libby Manuscript Collection, University of North Dakota, pp. 75-77. All subsequent quotes from Wiens are from this memoir.

[3] The strategic situation is well delineated in Frank and Cronin. See also A.A. Vandegrift, *Once a Marine* (New York: W.W. Norton, 1964) and Frank O. Hough, *Pearl Harbor to Guadalcanal: History of U. S. Marine Corps Operations in World War II* (Washington: Government Printing Office, 1958).

[4] Douglas Burtell, interview by Shoptaugh, October 12, 2007. The "British memos" referred to by Wiens are probably British press accounts of the Japanese victories in subduing Malaysia, Java and other objectives in Southeast Asia. In point of fact, as Frank shows, the average Japanese soldier was no more adept at jungle combat, no more comfortable in the harsh environment of the Solomon Islands, than the U.S. troops. "Jap" and "Japs" are now pejorative terms, but are quoted here as they were used in the sources.

[5] Alvin Tollefsrud, interview by Shoptaugh, August 28, 2007. The Marine sergeant's remarks are in Dennis Cline and Bob Michael, *Skeeter Beaters* (Elk River, Minnesota: DeForest Press, 2002), p. 188.

[6] Harry Wiens, "My Little Corner of the War," p. 76.

[7] Russell Schmoker interview conducted by staff of the State Historical Society of North Dakota, Bismarck, ND; Bennie Thornberg diary, entry for October 13, 1942. A copy of this diary was kindly provided by Shirley J. Olgierson, editor of the 164th Infantry Associations newsletters.

[8] Schmoker interview.

[9] Charles Walker, *Combat Officer: A Memoir of War in the South Pacific* (New York: Ballantine Books, 2004), p. 9-11. The Japanese air attack by 27 Mitsubishi G4M (Betty) bombers, escorted by 18 Japanese Zero fighters, was intended to damage Henderson and prevent American planes from interfering with the landing of Japanese reinforcements scheduled for the coming night. See Frank, p. 314. The comment about the American ships leaving before the unloading was completed is from John Stannard, *The Battle of Coffin Corner and Other Comments Concerning the Guadalcanal Campaign* (Gallatin, TN: privately published, 1982), p. 19.

[10] Information about Ken Foubert here and subsequent is taken from Shoptaugh's interview with William Tucker, September 16, 2007, who spoke about Ken's high school days, and a telephone interview with Colonel Richard Stevens (ret.), who noted that Ken "couldn't find a place to go to ground" when the bombing raid occurred and so moved into the palm grove. Stevens himself was able to squeeze into a shelter "built by Marine pioneers [i.e. engineers] but that was quickly filled." Additional information is from news clippings and documents supplied by Ken's nephew, Gary Foubert.

[11] The "job we had to do remark," in one form or another, was stated many times to the writer in interviews and conversations, by veterans of the 164th at their annual reunion in North Dakota in September 2007.

[12] Official notices of death gave families few details about what had really happened to their sons. The *Grand Forks Herald* reported that Foubert's family had received word that Ken had been "killed in action in the South Pacific, October 18 [sic]." The same issue also noted the death of Glenn Midgarden, from Company C, who died during an artillery attack on October 15. But again the War Department officially listed him as "killed in action October 18 in the southwest Pacific area." Private letters from other men in the regiment could well have told more: the Regimental S1 (Administration) Journal notes that the 164th regiment's HQ did not order "all mail frozen" until October 19th, by which time some letters could have gone out.

[13] Samuel Baglien, diary entry for October 13. There are actually two copies of Baglien's diary – an edited copy, widely distributed among 164th veterans in the 1960s and 1970s, and a complete copy made by Baglien's grandchildren. The second copy contains remarks he made about other officers, often unflattering. I appreciate the kindness of Balglien's family in providing with a copy of the unexpurgated diary, which will be added to the 164th Infantry Association Records at the University of North Dakota.

[14] Albert Wiest, interview by Shoptaugh, September 15, 2007, and John Paulson, interview by Paulson, October 20, 2007.

[15] John's Hagen's memories of his military service are recorded on 22 hours of tape deposited by his nephew at the Northwest Minnesota Historical Center, Minnesota State University Moorhead; hereafter cited as "John Hagen tapes."

[16] Here and below, "Mayville Man [Campbell] Recalls First Day on the Island of Guadalcanal as Deadly Lesson in War for Green ND Soldiers," *Grand Forks Herald*, August 2, 1992.

[17] Walker, *Combat Officer,* with additional detail in Walker's interview at the State Historical Society of North Dakota (hereafter SHSND).

[18] William Welander, interview by Shoptaugh, September 15, 2007, and Welander interview for State Historical Society of North Dakota, Bismarck.

[19] Rudolph Edwardson, interview by Shoptaugh, October 16, 2007.

[20] Baglien diary, entry for October 14,1942; tapes dictated by Edward Mulligan in the late 1990s and copied for Shoptaugh by Robert Dodd Jr.

[21] Vandegrift, p. 193.

[22] Frank Doe's memory was related to Shoptaugh by his son, Barry Doe, in a telephone interview, September 24, 2007. Frank Doe, now deceased, remained in the Army after the war, rising to the rank of major. He saved several documents from the regimental headquarters of the 164th which his son kindly shared with the 164th Infantry Association. When used these documents are cited as "Frank Doe Papers."

[23] Leckie, (here and below), pp. 98-101; Herbert C. Merillat, *Guadalcanal Remembered* (New York: Dodd Mead, 1982) p. 178. Merillat was part of the well-organized public relations system employed by the marines to publicize the efforts of the Corps in the Pacific. One method employed was the use of stories featuring individual marines and their exploits (what they called "Joe Blow stories"). This proved so successful at Guadalcanal that it was popularly (and wrongly) seen as a purely Marine victory. As we shall see, the 164th soldiers had their own ways of getting out some of their stories, but these generally reached only local audiences.

[24] William Bradford Huie, *Can Do!: The Story of the Seabees* (New York: E.P. Dutton, 1944), pp. 42-43. Seabees were civilian workers who were given a little training and organized into semi-military units (Construction Battalions, or CBs – Seabees) for construction of airfields, bases, and the like. They have never received the proper recognition for their contributions to the Pacific victory.

[25] Baglien diary entry for October 14, 1942; Leckie, p. 101.

Chapter 2: Hard Times
North Dakota: 1915-1940

Before they were soldiers, most of the men in the 164th Regiment were boys growing up in a very rural, less-than-ideal world. In the 1870s and 1880s, thousands of families came to Dakota Territory and established farmsteads under the terms of the 1862 Homestead Act. The land was all but given away by the Federal government, provided the recipients worked hard to make it productive. That was not an easy task. North Dakota came into existence just as the nation was making its dynamic and painful transition from an overwhelmingly agricultural nation into an industrial and increasingly urban giant. The population of cities grew explosively, until by 1920 more people lived in urban areas than in rural districts. Meanwhile, constant fluctuations in the price of grain made it ever more difficult for smaller farming operations to survive. Eric Sevareid, a future radio journalist who grew up in North Dakota, called wheat the "sole source" of his neighbors' livelihoods. Farmers, Sevareid said, "were never [wheat's] masters, but too frequently its victims." Wheat prices throughout the 1880s, for example, were often too low to justify the costs of growing and shipping the grain. Some did well, but a farmer who had borrowed money to get land or equipment found it enormously difficult to stay ahead of his creditors.[1]

Changes in personal and communal fortunes were sudden and wrenching in this era. During the 1890s, a depressed national economy made it almost impossible for the North Dakota state government to collect county land taxes. The state's operating budget was generally in arrears. Things brightened a bit as the new century began. The cities raised consumer demand for cereal grains, and the better prices for wheat allowed many farmers to pay down their debts. The population of the state grew steadily. World War I destroyed millions of acres of European crops and created an even greater demand for American food staples. So wheat profits rose handsomely, but many of the farmers had little chance to capitalize on this bounty when drought conditions lowered wheat yields between 1915 and 1919. By the end of the war, the state was entering what the state's premier historian called another "period of relative stagnation."

In the 1920s land indebtedness and inability to pay taxes placed many farms on the auction block. The population also began to fall as the birth rate dropped and "more [people] were leaving than were coming in." Having grown from frontier to statehood, North Dakota was by 1925 struggling to be more than what one historian called "a dependent hinterland [that was] producing wheat and cattle for outside markets and lacking [economic] opportunities for all of its people." When the Great Depression descended in the 1930s, conditions became even worse. Drought in the "dirty 'thirties" made the Plains a "dust bowl." There was so little water by 1934 that farmers killed thousands of pigs they could not afford to feed. Cows choked to death from dust in their lungs. In 1936, less than nine inches of rain fell on the state, summer temperatures reached up to 121 degrees Fahrenheit, and not even prairie grass grew. In one of the state's histories there is a photograph of a farmer standing in his wheat field. He is holding his arm about three feet above the ground, which was where his wheat should have been standing by then. The thin strands in the field were barely ten inches tall. No one was making any money. A memorable card sold in Edgeley offered this Christmas greeting: "Postage went up and wheat went down, collections are punk all over town." The rural population of North Dakota dropped seventeen percent in that decade.[2]

This was the land on which the men who joined the 164th Infantry Regiment grew up. They grew up on farms, mostly, knew what it meant to work long hours, and accepted at an early age that they had duties to perform. It didn't matter if it was threshing wheat, shocking grain bundles, repairing harrows, clerking in stores, making deliveries or any of a myriad other tasks, they worked day and night, in any conceivable kind of weather. There wasn't a farm boy who had not had a cow or horse stomp his foot, who hadn't felt the biting cold of the winter wind as he hauled and stacked firewood. To help the family purse they hired out with threshing crews at harvest time, found jobs, quit school and sometimes rode the rails to find work. The state paid a bounty of five cents for each gopher tail, so many of them learned to shoot by hunting gophers.

Rudy Edwardson began working when his father "rented" him to a neighbor. John Edwardson had a small farm outside Brantford, on land first cultivated in 1880 by Rudy's grandfather. Rudy began doing chores almost as soon as he could walk, small things at first, then just about everything: "milking, cleaning barns, hoeing, hauling hay, pitching bundles." By the time he was ten, he "farmed out . . . I had to work for a next door neighbor.

I had to clean out his damn chicken coop. They raised chickens and sold eggs. I had to pull up the roosts and clean them and calcify them every day. I've never wanted an egg since." He had no idea what the neighbor paid for this because the farmer gave the money directly to his dad.

The family badly needed that money because of Rudy's mother's health. In the 1920s, she needed surgery. The doctor who performed the operation "billed [Rudy's] dad twenty-eight hundred dollars," and he "had to mortgage the farm to pay it." The debt was still on the farm when the Great Depression made their situation even harder. Rudy needed to find more work. "The same farmer whose chicken coop I cleaned had two sons. They hired me to run the tractor and mow and rake, and then stack hay. I'd be on top of the stacks and it was hot, miserable work. I got paid about a dollar or two dollars a day." Despite this, the family lost the farm. Facing a life of low wages for backbreaking work, Rudy joined the National Guard in 1939, "to get away from home mostly."[3]

Bernard Wagner's father came from Iowa after World War I to buy a farm near the town of Marion. "Dad disliked humidity," Bernie recalled, "and found the North Dakota climate was better." Born in the hospital in Marion in 1921, Bernie attended the Greenland Consolidated School for five years and then went to Sanborn for the sixth grade. Like Edwardson, he worked on the farm and at jobs elsewhere: "the usual chores, milking, feeding hogs and chickens. I always got up at 4:00 a.m. and did chores before walking to school." Because his parents valued education so much, he was never taken out of school during planting or harvest. It was in Sanborn that he had his first brush with the National Guard. The high school basketball coach was a guardsman. He regularly recruited team members into the local guard company. In 1937, now part of the Sanborn High basketball team, young Wagner joined up. He was only sixteen, but one of his friends solemnly swore Bernie was eighteen and the company commander signed him up. "My parents were happy that I joined, because they liked the Guard, it was a neighborhood thing. They thought it would do me a world of good, and we got a dollar for each [weekly] drill." The money helped Wagner begin college, with three month's pay covering tuition for a quarter of classes.[4]

Melvin Bork's childhood was similar. He grew up on his father's farm outside Edgeley. Most of the farms in that part of Lamoure County were wheat farms; the Bork farm was no different. In good years the grain yields and a small herd of milk cows provided for the family. But money was generally tight, even more so once Melvin's mother arranged to rent a small room in Edgeley for each of the six children to use in the winter months while attending high school. "My parents wanted us to have a good education so we'd get more opportunities," Melvin later recalled. It was at the high school that he met some teachers and classmates who were members of the guard unit in Edgeley. He joined the unit in 1938, largely because "they were begging for people to join" and the extra money he got from the guard would help the family.[5]

Conditions were tough as well in Grand Forks where Bill Tucker grew up. His mother, Ruby Louise Tucker, raised Bill and his sister more or less on her own;

their father, a professional boxer, had left the family when Bill was four years old. Bill started taking odd jobs around Grand Forks while he was still in grade school. While attending Central High School, he worked after school and during weekends for the local branch of Western Union. "I got twelve and a half cents an hour from Western Union, so at the end of eight hours I had ninety-six cents." Tucker kept working, while his classmate Ken Foubert quit school to find full time employment.[6]

Life was not all about work, however. The boys growing up in North Dakota still found time to play baseball, hockey and other sports. They went swimming, read popular magazines and begged their parents for permission to go up for a ride when someone came around with a Jenny biplane. They scrimped to save a few bits of change and use it to send away for something they saw advertised in a newspaper. They made the most of what they had. John Paulson, who grew up south of Carrington, remembered that he and his cousin Miles Shelley, who lived nearby, would dream up a way to utilize any opportunity. "Miles would come up from his farm and we'd visit. If he had a nickel, or I did, then we both had a nickel, and we'd figure out something to do with it. We were real pals." Everyone listened to the radio. The whole nation was "in the doldrums" so they paid attention to the radio news. Many of them vividly remember their parents and neighbors arguing about the merits and shortcomings of the New Deal and of the disquieting stories of war in Europe and Asia. The movie houses cost about a dime, where, in addition to the features, you could see newsreels like the "March of Time," with its stories of gathering war clouds. Several remembered seeing newsreel footage of the "Rape of Nanking" in 1938, when tens of thousands of Chinese were massacred by Japanese soldiers. Few imagined that they would one day meet these same soldiers in battle.

Between their work obligations and their pastimes, these young men went to school, very often in a one-room schoolhouse. The schoolhouses were uniform in design and purpose. The classroom was invariably organized by age with the older students taking up the back rows of benches and desks. The teacher's desk at the front was flanked by an American flag and, often, the North Dakota state flag. Portraits of famous Americans adorned the side walls. Classes began with the pledge of allegiance and prayers were a standard part of the school day. Rote learning in everything from poetry to history to fundamentals of arithmetic was common. Most men interviewed for this book felt that they had learned a great deal and that it helped them better themselves throughout their lives.

If the boy did well in school, and his parents could afford to let him continue, then, like Edwardson and Bork, he went on to high school. Dennis Ferk was the son of a minister who was periodically assigned to different churches in North Dakota and South Dakota. "He had some skills in repairing churches, and even designed one in South Dakota that the congregation built," Ferk explained, "so we moved every few years." By the late 1930s, Ferk was living in Bismarck, where he attended a local business college. "I was learning secretarial skills, typing, that sort of thing. I also worked at a local restaurant, O'Brien's Café, to pay for the room I rented. I got twenty-five cents an hour." In 1939, Ferk joined Company A of the guard. "They were paying a dollar a week [for guard duty] and that helped with my room and tuition. I also used part of it to buy a radio, which cost me five dollars."[7]

But the higher education option wasn't always possible. Sometimes a family saw no need to send their boy to college or even high school. John Tuff's father was a well-off farmer in Pierce County, and for a time a member of the state senate. "They did well," John's widow Edith explained in 2007. "I remember that Eileen [John's mother] said that they bought a quarter section of land and paid for it in cash in just one year." The elder Tuff, Edith noted, expected all of his children to go to work once they reached their mid-teens. "John went to Banner School for grades one through eight, then he started high school in Barton. His father made him quit school at fifteen, and go to work at Pete Olson's farm, which was out there near their own. His father wanted him to go to work." John worked on the farms of his neighbors for several years, with the money he earned going to his father. John wanted to farm himself, and later worked in Montana to earn enough to buy his own land. He worked for a sugar beet farmer, who paid him in silver dollars. But before he could save enough, the growing international problems moved Congress in 1940 to pass a bill providing for military draft. John returned to North Dakota and signed up for the National Guard company in Rugby. "That way," Edith remembered, "he could serve with his friends."[8]

Men joined the guard companies in the 1930s for money and friendship. Russell Opat, a native of Minnesota, joined the Civilian Conservation Corps (CCC) after high school and learned to cook while working at a CCC camp at Grand Rapids, Minnesota. He found that he liked the military-style organization of the CCC and after a few years joined Company M in North Dakota because he knew some of the men in it. Several Native Americans who grew up on reservations in North Dakota joined the National Guard units. Woodrow Wilson Keeble, a full-blooded Sioux who was born on the Lake Traverse Reservation, spent most of his childhood in Wahpeton, North Dakota, where his mother had a job at the Indian School. Keeble, a standout athlete, joined Company I (Wahpeton) in the 1930s. He was destined to become one of the most decorated men in the regiment. James Fenelon, who grew up close to the Standing Rock Reservation, was the son of a full blooded Sioux on his mother's side and had a variety of French and Irish ancestors on his father's side. One of Fenelon's great-grandfathers, had been superintendent of the Devils Lake Sioux Agency back in the days of the Dakota Territory. Fenelon, like so many others, began working young, first by selling newspapers and then buying an old vehicle with a few friends to collect and sell scrap. "We even dug up cattle bones at places where they had shot their cows during the drought years. We dug up the bones and sold them to the scrap dealer." Fenelon joined the guard with some pals. He was enrolled at Devils Lake by Stafford Ordahl, a 2nd Lieutenant, who recorded Fenelon's age as 18, even though he knew Fenelon was only 15.[9]

The extra money was the biggest inducement for the young men to join in the 1930s. A 1980s survey of Dakota guardsmen conducted by two historians found that the "drill pay" was most often cited as the reason for joining in the depression years. "Few were motivated by any great desire to serve the country," said one man, who became a career Army officer. But camaraderie played a part as well. Robert Dodd, for example, grew up in Fargo, where Company B and the headquarters company were based. Dodd joined the HQ Company, lying about his age. As his son noted, Bob and some of his friends "were all technically underage . . . The Guard was in no way fooled by this, but was happy to ask no questions which might keep it from meeting its recruiting quotas." Dodd was attracted by the "much needed extra income" of the dollar per drill and fifteen dollars for the summer training. But he also admitted he joined because the "sharp uniform" looked great and gave a man "a little respect." Also, he could walk into the Bison Tavern and buy a beer, "no questions asked."[10]

Some were attracted by a sense of adventure. Albert Wiest, the career officer just mentioned, said that he joined because of restlessness. He and his twin brother, Robert, had been working as meat cutters, but Al was "too cantankerous" to settle into an ordinary job. "I had too much of an adventurous spirit. I played semi-pro ball around the state and was a pretty good ball player. I went down to Arkansas and played there. When times got rough in 1934, I joined the CCCs and spent a year in that. I was smart enough to be in college but there was no money, and frankly I was too lazy then to go to college." So he joined up. "They called this company [in Jamestown] the 'machine gun company' in 1933. They had .30 caliber Browning, water-cooled machine guns. Heavier machine guns and mortars came later. We were paid at the end of each drill, a silver dollar. That dollar a week was pretty big money then. It certainly wasn't a spirit of patriotism or anything like that that motivated us in those days." But as he spent more time in the guard, learning the techniques of the machine gunner, Wiest began to think he might enjoy being in the Army and "see new things." By 1940, he had advanced to 2nd Lieutenant and was committed to a military career.[11]

A love of music led others to the guard. Ralph Oehlke grew up in Enderlin, North Dakota, and decided to join the guard in 1939. Two of his friends joined to be in the regimental band, organized in nearby Lisbon. "We got three dollars a month from the guard and played with the band. I had played in my high school band, which was kind of an 'oompah' band. Through the guard, we got contacts to play some jazz at area school dances and so forth. You know, we were still just kids, pretty much, at that time." Many other men joined the band for the same reason. Leroy Busching, a self-taught trombone player, joined the guard because its members were often hired to play at dances – "we did a lot of one-nighters." Colonel LaRoy Baird, the commander of the 164th during the latter 1930s, frequently called out the band to play for state events. It was good public relations. Warrant Officer Gerald Wright, who led the band, was a music teacher in Lisbon who had a gift of persuasiveness tinged with a shade of blarney. Douglas Campbell, one of Wright's band members, called him "a promoter exceptional and a top notch recruiter." He scoured the state for talented musicians. He was so good at getting them to join that he could muster four dozen musicians for parades and concerts. U.S. Army regulations limited a regimental band to twenty-eight members, but Wright ignored this and at one point had a band of more than fifty players.[12]

Campbell recalled that Wright "had one persuasive pitch I heard him use on a number of occasions and that was, that after WWI, bandsmen would never [again] have to serve on the front lines doing litter bearing duty because many of the country's best musicians had been killed doing such duty during WWI, and that Congress had made certain this would not happen again." Yet, in addition to their playing and regular drill, Campbell and his fellow band members were taking "first aid training classes" from the regimental surgeon, who always told the men "make sure the first aid you give isn't the last aid." Why Campbell or anyone else never asked how they were supposed to administer first aid without being on the battlefield is a mystery. Only later would they learn the truth.[13]

The state guard was actually older than the state: Company A of the Dakota Territorial Militia had been formed in Bismarck in 1883, six years before North Dakota was admitted to the Union. North Dakota's guard units became part of a region's culture. Ranking officers of guard units were usually locally prominent businessmen or office-holders; the local guard company was often at the center of community life. In the 1930s David Ritchie, LaRoy Baird, Heber Edwards, and Earle Sarles, all active in state politics, served as either commander of the regiment or as adjutant general for the state. Edwards for example had a law degree, owned a thriving insurance business, had attended both the Army's infantry and command schools, and was trusted by both U.S. Senator William Langer and Governor John Moses.[14]

Each year, the U.S. Army parceled out equipment and weapons to the local armories, and contributed funds to each of the states for use in maintaining its guard companies. The equipment was invariably old stuff, Great War vintage rifles and uniforms, in the 1930s. North Dakota received an annual sum of thirty thousand dollars for this purpose during the 1920s, but that figure fell when the depression hit. In the 1930s there was also barely enough, when supplemented by state funds, to keep the system going. Most of that Federal money was used to pay the "drill fee" -- a dollar to each man for the weekly drills. An Army officer visited guard companies occasionally to inspect the troops, and once a year an Annual Federal Inspection was duly filed in Washington and the guard for each state.[15]

Every year as well, the North Dakota guard companies would meet at Camp Grafton, a 1500 acre patch of land in the vicinity of Devils Lake which was purchased by the state in 1904 and turned into a military reservation. Here the men took part in regimental level tactical exercises, honing themselves, as it were, for "the real thing." The quality of the training varied considerably. Many men who served in the North Dakota units before World War II later testified that the training they experienced at weekly drills was "poor" or "uninspired." Too often, many said, the company officers or sergeant relied on rote methods or simply read from manuals. Arthur Timboe, an officer then in the howitzer company at Devils Lake, laid some of the blame for this on lack of proper funds and reliance on old equipment. "Very few had any knowledge of administration or logistics. There was no adequate space for maneuvers or other operations." He also noted that there were never heavy machine guns or mortars for real tactical exercises. And while "unit training was stressed up to battalion level," there was "very little, not enough training in regimental size field maneuvers, or what to do if attacked or to attack." Albert Wiest pointed out the silliness of training only in fair weather – "it was unheard of to keep troops out overnight or in heavy inclement weather."[16]

Arthur Timboe was one of the best of the younger officers. So was Robert Hall, the future commander of the 3rd battalion. Hall's experience in World War I impressed men. When he described the conditions of the trenches in 1918 they paid attention. Samuel Baglien, another respected officer, was a barber who had joined the army in 1918, but was still in training when the conflict ended. Discharged in 1919, he returned to his native Hillsboro, North Dakota, and resumed his civilian occupation, joined the guard and slowly worked his way up, until by 1940, he held the rank of Major and was part of the regimental command. Men liked Baglien for treating everyone decently, from the lowliest private to the governor of the state.

There were other officers that they thought less of, some for having a "snobbish attitude" that small town residents recognized only too well. Charles Walker, a college student who joined the guard in 1937 and later received a commission as a second lieutenant, saw that the men respected officers like himself, who went through the same drills and exercises as the men. Other officers, those who just counted cadence or watched the drills, they despised. Walker decided that those "too old or incompetent," or "liked to lord it over the men," would fail at command if it ever came to combat.[17]

To be fair, many of the men in the ranks did not take their training very seriously either. Recalling the annual gatherings at Grafton, Douglas Campbell summarized his tactical exercises as "some pistol practicing on the range and some field training." He recalled that there was little attention given to serious war games and found that "the sand flies which could get in your eyes and in your mouth when playing [instruments] for parades" to be the most onerous part of the two weeks of training. At the camp the men lived in pyramidal tents erected over floorboard platforms. There were no field latrines, while the permanent ones had porcelain privies and showers. The mess was clean, the food good and ample, and about the only complaint made of these facilities was that there was "no hot water" until the later 1930s. Men, in fact, took to calling these two weeks the "summer camp" and the maneuvers as "glorified scout outings." With outdated equipment and few veterans of actual combat, the tactical exercises at Camp Grafton could do little to simulate real combat. The Army did not issue sufficient ammunition for serious machine gun practice until after war began in Europe in 1939. Don Hoffman remembered how excited this made the machine gun crews. "They were late in setting the targets up, so Al Wiest and his assistant gunner saw some birds over in a field and opened up on them. They got their asses chewed out good."[18]

Europe's war brought a new tone of seriousness in 1940. The encampment that year was at Fort Ripley, Minnesota, and was more ambitious. Four regiments from North Dakota, Minnesota, Wisconsin, and South Dakota formed an ad hoc division for field exercises held over three weeks. Dennis Ferk recalled the maneuvers at this one. His platoon made advances with some light tanks, "which we had never seen before." He got more practice with his Browning Automatic Rifle than ever before, but peacetime penny pinching still lingered. "We were very careful with ammunition and picked up our brass [cartridge] casings after practice. Nothing was wasted." "We were still playing at soldiers," another man commented. It was at Ripley, though, that men for the first time heard rumors that they would soon become real soldiers. President Roosevelt declared a state of "unlimited national emergency" in May 1940, just after Hitler was launching his all-out attack on France. A bill to authorize a peacetime draft was introduced in Congress and it soon became clear that it would pass. The nation was finally taking some serious steps for proper defense. Bob Sanders, a member of the regimental band, remembered that "in the summer of 1940 when we were at summer camp, [we] got the word that very likely we were going to be called up for a year's active duty and that probably we would end up somewhere in Louisiana and that we would be called out in September." The U.S. Army did call up several guard units right away, but delays in new camp construction forced the mobilization orders for other units to be delayed. First the 164th was told that it would mobilize in mid-October. Then the date was postponed again until the end of the year.[19]

The Guard in North Dakota now worked feverishly to get more men to fill out their rosters. When Congress passed the draft bill, a number of men decided to sign up and serve with their neighbors and buddies. Douglas Burtell was one of the men who joined up at this time. He grew up in Casselton, where his dad had a job as a mechanic in one of the local garages. Doug lived about a mile from his grade school. "We ran home from school at noon for our meal, and ran back, did all that in an hour." Doug went to work at about the age of thirteen, at the garage where his dad repaired vehicles. "I wasn't able to play football until my junior year, because [before that] I had to be down at the garage at four o'clock, scrubbing floors, washing cars, greasing cars, driving the wrecker – I didn't have a driver's license. I got my social security card when I was fourteen, in 1938. Then when I was fifteen I delivered a truck to a farmer. I got that job hauling grain [from him] for a dollar a day. Later, the farmer fired one of the bundle haulers who got drunk, so then they put me on a bundle wagon and I hauled bundles for a dollar a day. That was hard work, you got up at four or five in the morning..." Doug's brother, Sidney, worked at many of the same jobs on other farms. He joined the navy in 1937. In 1940 Doug joined the headquarters company of the National Guard. "Some of the guys that were in the guard, like Bernard Starkenberg and others, would come out to Casselton. Some were chasing girls, others had family and friends there. Starkenberg and these guys talked me into joining. I was sixteen years and eight months old. That was in December 1940. When we were sworn in to Federal service in January, that old sergeant stamped out our World War type dog tags with a little stamper and a wooden hammer."[20]

John Hagen's decision to join the 3rd battalion headquarters was similar. Hagen lived in Moorhead, Minnesota, just across the border from Fargo, North Dakota. He was the eldest son of Olaf Hagen, one of the most prominent surgeons in Minnesota and a major player in the state's Republican Party. In 1940, Dr. Hagen ran for the party nomination for the U.S. Senate and came close to winning. His wife dead for a decade, always busy with his practice and politics, Hagen relied on housekeepers to raise John and his siblings. John was an indifferent student at school, tried several colleges and dropped out of all of them and by his own admission, preferred "pooping around" with his friends at the local taverns to anything serious. One day in the fall of 1940, John came home and told his dad that, with the war going on in Europe, he and his pal "got the bright idea that these Canadians needed some hot pilots" and they were going to go up to Manitoba and join the Commonwealth Air Training program in order to become fighter pilots for the Royal Air Force.

Dr. Hagen was horror-struck by the idea of his son going to war, the more so because he had run his senate campaign on the platform that the U.S. should be "strictly neutral" in the European war. He had no intention of letting John carry out his plan and he quickly found an ally who could help him keep John close to home. This was Bill Boyd. A native of Fargo, Boyd was in Company B of the guard. Boyd knew John Hagen: "I was friends with a lot of friends that he had, Ozzie Frederickson and Jim Bruso, and that group. That is really how I got acquainted with John. I remember John was going up to Canada with two or three others guys to try to get into the Canadian Air Service, which I talked them out of." By this he meant he talked John into joining the guard. Several of John's other pals took him to a bar in Fargo and then, after several beers, walked him over to the Armory to sign up. As Hagen later explained, "they had already made everything out on the forms the best they could. I figured to be a good fellow I could sign up for a while. They said I 'could get out anytime' and all of this." Bill Boyd said "let's just say we softened him up a little, we took advantage of his youth and innocence." So, with a little nudging from his friends, John joined the guard. He never expected this would put him into a war.[21]

However, there were those who went in with their eyes wide open. Edwin Kjelstrom, a twenty-three year old salesman at one of the men's clothing stores in Rugby, joined Company D in late October. He remembered how Captain William Mjogdalen, the company's new commanding officer, made a point of telling each new recruit that "it was possible that the U.S. would be going to war" during their year of service. Up in Edgeley, the men best trained in first aid were reorganized into a new detachment. Melvin Bork and some of the others were transferred to other duties with the 3rd battalion headquarters. Some of the infantry platoons were converted to the artillery. In recognition of the growing power of armored vehicles, the old howitzer company was converted into an anti-tank company. This was placed at Harvey and staffed with some experienced officers and men. Ralph Oehlke soon found himself attached to this unit, which at first had almost no weapons. Finally, LaRoy Baird and Heber Edwards decided they were too old to take active service in a war, so Earle Sarles was promoted and designated to command the regiment as it entered the Army.[22]

One of the last steps taken before the regiment left was the printing and distribution of a "Historical and Pictorial Review" of the regiment. This large volume is a rather bittersweet confection. Filled with carefully arranged photographs of all the officers, all the men and detachments, and a long essay on the unit's history, it looks like a high school yearbook, a special bon voyage for a graduating class. A special statement addressed "to parents" asked "would you prefer that [your sons] serve in the Guard under officers who know your boy and love him, and who will when the service is terminated come home and answer to you for any dereliction of duty, or would you prefer he serve under strange officers who do not know him and whom you will never see if your boys are abused and neglected?"

In December the Army notified the regiment that it would be mobilized in early February 1941. The companies and detachments reported to their armories in late January for inspections and physicals given by Army physicians. A number of the older men did not pass these physicals, so some were discharged and everyone else was put through strenuous drills to get them into better shape. Doug Burtell vividly remembered those last weeks. "We moved into the Armory in Fargo. I was put in Headquarters Company. We had all World War I equipment, wrap leggings, Springfield rifles, and we trained in Fargo in 25 degree below zero weather, out in the park. Stood retreat every night uptown in front of Penney's." Already a talented artist, Burtell made some sketches of these days. One shows the men holding retreat on Broadway, another depicts him and another soldier trying to sleep on their cots in the armory. It was so cold, he noted, that the men spread their overcoats on top of the two heavy wool blankets they had been issued. These scenes were repeated at the other armories where the men waited while the trains were organized to take the regiment south. The average age of the regiment was about twenty-three, and the men were excited; many of them had never left the state before and almost all of them were looking at this as a great adventure.

Americans had accepted the need for an enlarged Army, but polls taken throughout 1940 had shown that the American public was overwhelmingly against the U.S. entering into the wars raging in Europe and China. Congressmen and senators from every state had pledged that they would only vote for war if America was attacked. When President Roosevelt ran for re-election in November, he promised the voters "your boys are not going to be sent into any foreign wars." So, as the history of the state guard noted, the men amused themselves while waiting by singing a new popular song, "Goodbye Dear, I'll be Back in a Year."[23]

And yet, America was already in the war, at least economically. When the war began in 1939 with the German invasion of Poland, Roosevelt had called Congress into special session and asked for a revision of the prevailing neutrality laws. Change the laws, he argued, so that American companies would be permitted to sell goods, including weapons, to any nation on a "cash and carry" basis. The cash provision eliminated the credit problems that had been an issue in the Great War, while the carry provision would keep American ships out of the war zones – and the sights of German submarines. Roosevelt said that this would improve the nation's chances of remaining neutral. A number of congressmen had fought against Roosevelt's proposal, including North Dakota's Senator Gerald Nye, who had helped write the neutrality laws. Nye called this "cash and carry" bill a first step toward war. He pointed out that Germany had no large navy to protect merchant ships, so cash-and-carry would patently favor Britain and France.

But the public blamed Germany for starting the war in Europe, and a slim majority in Congress favored the idea of selling to the Allies. The cash-and-carry amendment passed. The Allies bought about a quarter-billion dollars worth of arms from American companies within the first month. These sales included Lockheed Hudson bombers, sold to France, and delivered by flying them to Pembina, North Dakota and physically hauling them across the Canadian border. As the year drew on, Britain began running out of money to buy weapons, so Roosevelt asked legal advisors to see if America could "lend" them aid. This "lend-lease" provision passed Congress in 1941.[24]

By then the 164th would be training in Louisiana, and several of its men would come to think that a war against Germany was possible. But almost none gave a thought to war against Japan. Japan was dismissed as too weak to challenge America. None could know that, while they waited for their transportation to Louisiana in February 1941, officials at the State Department in Washington were pondering a report from their ambassador in Japan, who wrote that "he has heard from many sources, including a Japanese source, that in the event of trouble breaking out between the U.S. and Japan, the Japanese intend to make a surprise attack against Pearl Harbor with all their strength and employing all their equipment." The State Department shared the report with the Army and Navy Departments. Their leaders dismissed it as too fantastic to be taken seriously.[25]

(Endnotes)

[1] Elwyn B. Robinson, *History of North Dakota,* pp. 367-74.

[2] Theodore B. Jelliff and D. Jerome Tweton, *North Dakota: The Heritage of a People* (Fargo : North Dakota Institute for Regional Studies, 1976), pp 155-56, for the drought conditions, farmer's photograph and Edgeley Christmas card; Robertson, p. 374, for the "dependent hinterland" observation.

[3] Shoptaugh interview with Rudolph Edwardson, October 16, 2007.

[4] Bernard Wagner, interview with Barnes County Historical Museum, August 20, 2000.

[5] Shoptaugh's interview with Melvin Bork, August 31, 2007.

[6] Shoptaugh interview with Tucker, September 16, 2007.

[7] Shoptaugh interview with Dennis Ferk, September 16, 2007.

[8] Shoptaugh interview with Edith Tuff, August 6, 2007.

[9] Shoptaugh telephone interview with James Fenelon, April 2007.

[10] Robert Dodd Jr., *Once a Soldier: Robert Dodd, the 164th Infantry and the War in the Pacific* (privately printed, 2001) pp 15-31.

[11] Shoptaugh interview with Albert Wiest, September 15, 2007.

[12] Ralph Oehlke, interview for Barnes County Historical Museum, March 23 2004; Douglas Campbell's written reminiscences in the *164th Infantry, IInd Island Command, 294th A.G.F. Band Reunion* booklet, October 1985. Leroy Busching memory from article in *Ransom County Gazette and Enterprise,* January 21, 1987; Gerald Wright information from narrative in the *164th Infantry, IInd Island Command, 294th A.G.F. Band Reunion* booklet, October 1983.

[13] The memory about Wright assuring men that they would not become litter carriers is from Douglas Campbell.

[14] Here and below, the history of the Guard before World War II is drawn from Jerry Cooper and Glenn Smith, *Citizens as Soldiers: A History of the North Dakota National Guard* (Fargo: North Dakota Institute for Regional Studies,1986), esp, pp. 218-58. See also *Historical and Pictorial Review: National Guard of the State of North Dakota* (private printing, 1940), a special volume published at the time of the unit's mobilization for Federal service. The Guard's ties to Federal service were based on the 1792 Militia Act, and a 1903 law that modernized some procedures.

[15] Robert Bruce Sligh, *The National Guard and National Defense: The Mobilization of the Guard in WWII* (New York: Praeger Press, 1992), is excellent for detail on the Federal-State Guard relations and arrangements in the 1930s.

[16] Timboe comments from Cooper and Smith, and Wiest remarks from his interview with Shoptaugh.

[17] Timboe and Baglien information from Cooper and Smith, and from Shoptaugh correspondence with Mary Fran Riggs, Timboe's daughter; SHSND interview of Charles Walker.

[18] Campbell comments from his memoir in the *164th Infantry, IInd Island Command, 294th A.G.F. Band Reunion* booklet, October 1985, and additional details from short interviews with 164th veterans that were videotaped and preserved at the National Guard headquarters in Bismarck. See also Shoptaugh interview with Don Hoffman, September 21, 2007. Discipline was a rather haphazard affair in the 1930s. After joining Company K in 1939, Horace White went to the west coast to get work. He failed to make enough to pay for a train ticket back, so he hopped "a freight train to get home for the Grafton exercises: "I rode as a railroad bum." He was too late for the camp. Listed as AWOL he went before Captain Frank Richards for discipline. "He just gave me a good bawling out and told me not to let that happen again. He felt sorry for me." Richard would command one of the regiment's battalions on Guadalcanal. Howard Woodrow White, interview by James Fenelon, September 20, 1996, transcript, p. 2. I owe a special debt of gratitude to Mr. Fenelon, former editor of the *164th Infantry News*, for sharing copies of the interviews he conducted in the 1990s.

[19] Robert Sanders memoir taken from tapes he dictated in the early 1980s which were reproduced in the *164th Infantry, IInd Island Command, 294th A.G.F. Band Reunion* booklet, October 1985. For detailed information of the expansion of the U.S. Army in 1940, see Mark Skinner Watson, *The War Department: Chief of Staff: Prewar Plans and Operations* (Washington DC: United States Army, 1950), pp. 183ff. This volume is part of the series *The United States Army in World War II,* an indispensible source for understanding the American war experience.

[20] Shoptaugh interview with Douglas Burtell, October 12, 2007.

[21] Information on John Hagen from his recorded tapes and from Shoptaugh, "Never Raised to be a Soldier," *North Dakota History,* vol. 66 (winter and spring issues, 1999).

[22] Previously noted interviews with Bork, Ferk, and Oehlke; Edwin Kjelstrom memoir from *164th Infantry, IInd Island Command, 294th A.G.F. Band Reunion* booklet, October 1983; Cooper and Smith, *Citizens as Soldiers,* p. 266.

[23] Burtell interview; Cooper and Smith, *Citizens as Soldiers,* p. 266.

[24] Franlin D.Roosevelt, "Repeal the Arms Embargo," *Vital Speeches of the Day*, October 1, 1939, pp. 738-41; Shoptaugh, "Borderline Neutrality," *North Dakota History,* vol. 60 (Spring, 1992).

[25] Hiroyuki Agawa, *The Reluctant Admiral: Yamamoto and the Imperial Navy* (1982), p. 219.

Chapter 3: In the Army Now
Louisiana: 1941

The men of the 164th were sworn in for Federal service on February 10, 1941, but it took another two weeks before they actually set off for Louisiana. They were part of a new American Army being born in 1941. Only 190,000 men had been on the Army's rolls in 1939. But with Germany overrunning much of Europe, the American Congress had finally moved to enlarge its military. The draft would bring a million men to the army. Guard units that were activated added another quarter million. The Navy and Air Corps were given the green light to grow similarly. The nation was awakening to the dangers it faced, but it would take time to assemble these new forces and train the men.

More than any other person, General George Marshall could be called the father of this nascent army. Roosevelt had selected Marshall in 1939 to become his Army Chief of Staff. A soldier with an impeccable record, a keen mind, and reputation for accepting nothing less than excellence, Marshall was the man who convinced the president in May 1940 that the American military must be expanded without further delay. Pointing to collapse of France before Germany's onslaught, listing the dangers that the United States must now face in both the Atlantic and the Pacific, Marshall told Roosevelt that he could not guarantee America's ability to defend itself without more men, ships and planes, without better weapons and better equipment. Roosevelt yielded to Marshall's pressure, Congress voted for an emergency defense appropriation, and the draft was authorized. A new army was conceived, but its gestation was going to be difficult.

It was Marshall who found the officers who could take raw draftees and half-trained guardsmen and shape them into a real military force. One of these officers was General Leslie McNair. Marshall plucked McNair from his command of the War College at Fort Leavenworth, Kansas, and put him to work designing new training procedures that would be used to create warriors out of the young recruits. McNair, in turn, set his staff to work and together they churned out a carload of new training manuals in a matter of months. Marshall was so pleased with McNair's work that he promoted him first to the position of Chief of Staff of the General Headquarters, then, in 1942, to Commanding General of all Army Ground Forces.

Meanwhile, Army engineers and civilian contractors built dozens of new training facilities. A major named Leslie Groves, head engineer for the building of the Pentagon and the future military director of the atomic bomb project, was chosen to ramrod the completion of over forty camps. One of these was Camp Claiborne, built outside the town of Alexandria in central Louisiana. On paper, Claiborne met all of the specific requirements that the military wanted for the proper training and housing of its new army: it had thousands of acres of varied terrain for tactical exercises and was close to rail lines and a town big enough to provide the soldiers with at least some place to go when off duty. Claiborne was built quickly and as a result had many defects, as the 164th would discover.[1]

As the day of the regiment's departure approached, the host towns for the companies prepared farewell ceremonies. Dances were held, mayors gave speeches, choirs sang, and churches held services. At most departure sites, veterans of the First World War and members of the local American Legions acted as honor guards as the companies marched to the stations. Governor John Moses came to Fargo see off Company B and the Headquarters troops. The farewells were both poignant and extravagant. Bill Hagen, a member of B Company, remembered how Carl Eggen, owner the Bluebird Grocery in Fargo, ran back to his store to gather up a parting gift. "He came back to the train depot to say goodbye and gave us bags of goodies and candy. They were good." John Hagen (no relation) remembered how he and his pals boarded the train "all in dress uniform, our b[arracks] bags filled with clean underwear and brown bottles of whiskey." Others, like John Paulson and Alvin Tollefsrud, spent the trip thinking about their families. Doug Burtell, who wouldn't be seventeen for another two months, remembered the "great sense of excitement" that permeated his drawing car.[2]

Most men rode the rails, but one unit drove to Louisiana. The regiment's trucks were transported by members of the Service Company, along with Doctor George Schatz and some men from the Medical Detachment. Gerald Sanderson, a medic, acted as a relief driver. Sandy was a native of Willow City. He had joined the regiment in December 1940 "to get my year of Federal service over with." During his time with the Medical Detachment in Bottineau, he had learned quite a bit about first aid, but had "never had my hands on a Springfield" rifle. Now, as the trucks started south on U.S. Highway 75, he was given a rifle to help guard the trucks at stops. He also discovered that maintaining the "proper space" in the convoy was something of a challenge, "especially for us greenhorns." The trip took seven days, the convoy stopping each night to sleep in some type of public building. Only in Missouri did they have any excitement. That occurred near Joplin when a farmer driving an egg truck tried to cut in front of one of the big Dodge trucks, slammed into its rear axle, and scattered broken eggs all over the pavement. After getting the truck repaired the convoy went on to Louisiana.[3]

Most of the rail journeys took four days, the trains stopping only to take on fuel and food and give the men brief periods to stretch their legs. When they arrived at the end of the line in Alexandria, Louisiana, the soldiers were taken by trucks to Claiborne. They took one look at the camp and were immediately disappointed. Most of the wooden buildings were little more than half-way built. The men were to be housed in pyramidal tents over wooden frames with plywood floors. The tents were arranged in endless blocks separated by dirt lanes. There were no signs marking anything as yet, and everyone got lost trying to find their way around. There were also no stoves for heat, so at first the men just covered up as best they could.[4]

Douglas Campbell decided that after a few days he and his buddies were ready to jump a train for home. "Soil was being moved about" the camp, Campbell recalled, "and it smelled like hog manure. Our disappointment [with the camp] was very great. The heavy four buckle overshoes we had hoped to dispense with became our most prized possessions. The heavy winter clothing which had sufficed during sub zero weather back in North Dakota didn't serve the purpose against the damp penetrating coldness of the sunny South. Our entire area was covered with ankle deep mud." Bill Tucker, a private with Company M, hated humidity: "you took your clothes off at night and hung them up and they were very damp in the morning." A lot of the men were sick in the early weeks, mostly from colds, sore throats, and sinus problems. When gas heaters finally arrived for the tents, Campbell and his friends "became resigned that this was going to be our home for the next year and we might as well make the most of it."[5]

Camp Claiborne was surrounded by boggy land that was crisscrossed with creeks and narrow dirt tracks. The red clay soil became a thick gumbo after even a short rain -- and it rained a lot. Ossie Frederickson, one of the veteran sergeants in the headquarters company, described the weather in early March: "It has been raining all day and the ground is a hole of mud and corruption." The unfamiliar terrain fostered other troubles. While on hikes, men took breaks in what appeared to be weeds and ended up severe rashes from poison oak. They were tormented by mosquitoes, which were at least a known quantity, but chiggers, ticks and cockroaches were entirely new to them. So were some of the swamp-dwelling reptiles. One day Bill Welander took part in a hike during which his platoon came to one of the creeks. Welander thought they might be able to jump the creek. "I was standing there looking at the depth of the water and all of sudden I'm looking into the eyes of a blue racer snake. I cleared the creek without touching the mud on the other side."[6]

After receiving letters about the conditions, some of the men's families complained to Governor Moses that their sons' health was being endangered. Dr. Olaf Hagen, John Hagen's father, went to Louisiana to see his son, looked around and satisfied himself that there was no threat of epidemics. He returned home and reassured the local press that the North Dakota boys were "just fine." John Hagen's frequent letters to his dad paint a good portrait of life at Claiborne. In March he wrote to thank Dr. Hagen for sending him an 8mm movie camera, which he used to shoot footage, which still exists, of Claiborne and Alexandria and his buddies. He also was pleased to note that the headquarters company had "just got 13 new [M1] automatic rifles, the newest thing on the U.S. market, 8 shots in rapid succession." Soon after, he reported that many of the locals were not very friendly. Having gone to visit "a beautiful brunette" who had gone to college with his sister, Hagen was offended when she asked him to wear civilian clothes when he called again. "Otherwise the neighbors would know I'm a Yankee."

Hostility from the locals went beyond regional prejudices. Soldiers were not yet seen as saviors. Don Hoffman remembered that the first time he went to Alexandria he saw a sign at the municipal swimming pool that said "no dogs or soldiers." Local bars refused service to soldiers and local people refused to talk to them. About the only people in town who were happy to see them were men who regimental historians termed "hucksters out to make some quick dollars from the soldiers." Finally, one of the officers, possibly someone from the headquarters of General Ellard Walsh, commander of the 34th Division to which the 164th was attached, told the mayor of Alexandria to remove the offensive signs before they triggered violence. The signs came down but a lot of Alexandria citizens remained unfriendly until after Pearl Harbor.[7]

Basic training began immediately. At first it was mostly calisthenics, marches and basic weapons training. Then the platoons and companies moved on to tactical exercises: assaulting entrenched positions; support fire; fire and movement; the advantages, and dangers, of defilade positions; map orientations; terrain exercises; smoke exercises; anti-gas training; and field communications. The exercises drilled the men to take up their positions and perform their moves automatically. They learned to fight in larger formations, taking part in multi-company, multi-battalion and ultimately-multi-regimental maneuvers.

Poor weather interfered with the some exercises and equipment shortages hampered others. Targets for rifle and machine gun practice were late in arriving. The machine gun crews were able to practice with the standard water-cooled .30 caliber guns, but few of the heavy .50 caliber guns were available. Outdated tactics did not help either. Al Wiest noted that 'indirect fire' was overemphasized: "We spent too much time on that, showing men how to elevate the guns and bring fire down on a target that was out of their view. Mortars and light artillery had been developed for that and we never really used it in the Pacific." At first mortars were in short supply. The mortar crews had no real tubes to practice with. "They had us cut limbs and shape them into mortar tubes to use for practice drill," Don Hoffman commented. "We did a good job making the limbs match the specs for the mortar, so the lieutenant bought us a case of beer." They spent weeks setting up and sighting the wooden mortars. Ralph Oehlke was part of the anti-tank company, which likewise used wooden mockups of 37mm guns. "We put wooden trailers on pipes and practiced loading them with old soup cans. We finally got just one real 37mm gun, and that was in December 1941."[8]

In the last part of the summer, the entire 34th Division took part in the "Louisiana Maneuvers," where two armies maneuvered against one another in complex actions. The men of the 164th took part with the 34th in the "Blue" Army, commandeered by Lieutenant General Walter Krueger. Most men generally remember these maneuvers as a "hot, miserable time" in which they marched long hours on dusty roads. Doug Burtell, taking part with a scouting force, was shocked by the "dozens of [cavalry] horses that were dead along the roads." The men saw lots of trucks marked with paint as 'tanks' or 'armored cars' and airplanes drop flour sacks to simulate bomb attacks. Bill Boyd, now a 2nd Lieutenant, was a judge during a battle and enjoyed watching the higher ranking officers argue over the judgment calls. Harold Larson of Company C was surprised when he heard that General Krueger had complimented the "guard troops" on their performance. Regular officers seldom complimented guard units for anything.

The Louisiana maneuvers influenced important changes in the new army. An official Army history of the maneuvers notes that "most of the forty-two division, corps, and army commanders who took part in the GHQ maneuvers were either relieved or reassigned." Only eleven of the forty-two generals went on to hold "significant combat commands" in World War II. Walsh, the 34th Division commander, was one of the men reassigned. His replacement, Russell Hartle, was a regular army officer of thirty years experience. Walsh subsequently claimed that the regular Army was prejudiced against National Guard officers.⁹

This was the first experience for most of the men in a regular army-national guard split that surfaced throughout World War II. By the fall of 1941 Colonel Sarles was complaining that regular Army officers ignored most of his suggestions. Guard officers most resented the Army's West Point graduates who they saw as members of a club who looked after their own. The West Point men were quickly dubbed as the "ring-knockers." But in many ways, the regular army-guard divide was a clash of *two* exclusive clubs. The Guard had its own cliques. Wendell Wichmann, a North Dakota native, learned the extent of guard clannishness when he transferred to the 164th. Holding a reserve commission, Wichmann was alerted for active duty in mid-1941 and joined the regiment in September. "People don't realize now how closed the ranks of a guard could be then. I got to Claiborne and they sort of said 'What in the world are *you* doing here?' A reserve officer coming in to a National Guard unit was not exactly welcome." Fortunately, he was assigned to Company D and John Tuff was his sergeant in the 3rd platoon. "John was a heck of a nice guy and Bill Considine, the battalion commander, took a liking to me."¹⁰

Despite so many equipment shortages, the men did receive the army's new rifle. The semi-automatic Garand M1 would be recognized as the definitive American infantry weapon of the Second World War. The earliest models of the M1 were produced by the Springfield Armory and some of these early M1s were beautiful weapons, built with walnut stocks. Dennis Ferk had the good fortune to be issued one of these when he was at Claiborne. He polished the weapon repeatedly until "the wood shone like it had been varnished; I loved that rifle." Bill Welander also became proficient with the M1. One day, Welander's company commander called for him to come to headquarters and "bring my rifle." I got there and there was a colonel there from Washington. Welander's commander ordered him to do the manual at arms for the visiting officer. "So I did it and at the end of it the colonel walked over and said 'Thank you.'" The colonel was Dwight Eisenhower.[11]

  The guardsmen slowly became real soldiers. The more lackadaisical ways of the past began to fade. Soon after the men got to Claiborne, some of the Service Company sergeants "borrowed" Colonel Sarles' staff car for a quick run into Alexandria. Sarles found it and was content to just admonish them, saying "next time get written permission!" By the fall, aware of the many Guard commanders who had been replaced, he was less forgiving of such high jinks. He cooperated in culling out some of the older noncoms and helped pick the most promising enlisted men for OCS courses, while others were raised to non-commissioned ranks. This included men like Rudy Edwardson, John Paulson, his cousin Miles Shelley, Bernie Wagner and several others. Ed Kjelstrom, a rifleman and part of the band, was promoted and ordered to "instruct the rest of the men in this new rifle. It was a very difficult time for me. I was not used to giving instructions and I was very shy. I began to know what it was like to be away from home and unable to say 'I quit.' Life had changed for me."[12]

After the Louisiana maneuvers, the 34th Division was streamlined. The 34th had four regiments in the old World War I style. General Marshall's intention was to create fully motorized, three-regiment divisions that would reflect the speed and power of the mechanized forces in Europe. Before late 1941, he lacked the vehicles to fully equip more than few divisions, but he proceeded to reduce the four regiment divisions in size. But what would happen to excised regiments? Around the 164th headquarters units, rumor had it that it would be the North Dakotans who would be "expelled" from the 34th, and a sense of incipient rejection began to filter down to the companies and platoons. Added to this frustration was the awareness that Congress was planning to extend the one-year service of the men. The Army could not afford to let its trained soldiers go after one year.

Few today can appreciate the anger that arose in America from this projected extension of the draft. Groups of mothers organized to lobby against it, newspapers condemned it as a 'breach of faith' and angry soldiers who expected to go home in 1942 began scrawling OHIO on the walls of barracks – "Over the Hill in October." The men of the 164th were affected as much as any unit. "What do you think the outcome of this bill before Congress will be," John Hagen wrote his dad. "Do you think that we will be kept down here more than a year? How about a little information on that?" 'They shouldn't have misled us when they mobilized us' was a common sentiment in the interviews of 164th veterans.

Without doubt the draft issue affected morale among all the troops. Hilton Railey, a reporter for the

New York Times, travelled through the army's camps in the summer of 1941 talking to officers and enlisted men by the hundreds. He compiled a two hundred page report on "morale in the U.S. Army" that he shared with the War Department. His basic finding was that, without a war, most men thought there was little point in their being in the service. They felt that they were being treated "like a football team in training but without a schedule of games." Railey also believed that soldiers doubted the ability of the officer corps to lead them into battle. Some of Railey's most shocking examples came from his visit to Alexandria, where one night he saw soldiers in "a carnival of drunkenness" openly insult the officers in the establishment. Marshall read Railey's report and was annoyed, but he needed this phase of army expansion to be a success. So he rejected a suggestion to demobilize many of the guard units, made some improvements in camp conditions, and approved a plan to discharge older soldiers who were having the hardest time adjusting to their new lives.[13]

The 164th continued its training. The men grumbled when the draft extension passed Congress by a hair, keeping most of them in Uncle Sam's hands for another eighteen months. Facing the likelihood that they would be removed from the 34th division, and wondering what the future held for them, the men found solace in the standard routines of camp life. They went on further training exercises, repaired the duckboard sidewalks, and laid sod to create some more green space in the camp. They took in movies at the camp, went to Alexandria, read, played cards, or sat around in seemingly endless "bull sessions." Every company had formed its own sports teams, and from boxing to baseball, competition and betting on the contests was heavy. Some men received short furloughs to go home and visit family or take off for New Orleans to chase girls. When Art Timboe was promoted, he celebrated by marrying Catherine Baker, a Devils Lake teacher that he had courted before leaving for the South. Naturally, the band played at the wedding reception. Timboe would be commanding one of the battalions overseas when his daughter, Mary, was born in August 1942.

On the weekends the men rushed to Alexandria and the numerous bars. Alcohol was a traditional lubricant for the military machine. When Alvin Tollefsrud's company mustered for a march after several of its members returned from a long boozy two-day leave in town, Tollefsrud noted that "Colonel Sarles saw us and yelled 'Here come those drunks of Company L!'" Bob Sanders, a member of the band, had the job of waking the band members. "I would go bang on the doors [of the tents] and I was greeted by a fellow saying 'come in and turn the gas stove on' or 'turn the light on as you leave' or sometimes I was met with a volley of army shoes. I also had to wake Gerald Wright up each morning. I would go in, turn the stove on, and light the light . . . he would put an arm down on the floor and haul up a jug of whiskey and pour about one-third or one-half glassful of that and gulp it down and then he would offer me one and I would turn it down." Doug Burtell remembered one old hand who got dead drunk each weekend. "One time while he was drunk a few guys went out to the road by the camp where there was a dead pig that had been run over by a truck. They got that muddy, bloody pig and stuffed it in his cot with him and got his arm around the pig." Their commander had a fit and called an assembly to find out who did it. "Naturally no one confessed so he made a bunch of us bury the pig."[14]

It wasn't easy to keep men away from Alexandria's bars and fleshpots. At one point during the summer, most of the 3rd Battalion personnel were quarantined at Claiborne after a couple of smallpox cases were discovered. A number of the battalion's sergeants led a bit of a revolt against the isolation order, led a "breakout" and took a bevy of men to town. They hit the bars and went to a local baseball game. Sarles broke most of the sergeants to corporals and ordered the battalion punished with "can duty." "We collected beer cans from the PX and crushed them for disposal," Les Aldrich recalled. "We stomped on the cans in the latrines and cleaned the latrines, too. They made up a song to make fun of us, singing that we'd 'win the war by crushing beer cans on the shithouse floor.'"[15]

In Washington, George Marshall had bigger worries than unruly sergeants. Germany had attacked Soviet Union in June of 1941 and by September, Hitler's armies were closing in on the capitol in Moscow. Few expected Russia to stave off defeat. German submarines and American destroyers had already faced off in a couple of incidents in the Atlantic. Japan, meanwhile, had taken advantage of the German defeat of France to occupy bases in French Indochina. President Roosevelt reacted to this by cutting off all oil sales to Japan which he hoped would force the Japanese to negotiate a settlement with the United States. But the Japanese Navy had other ideas. Admiral Yamamoto, the commander of the Japanese high seas fleet, insisted that a surprise attack on the American fleet at Pearl Harbor offered the only possibility of victory over the United States. With some reluctance, the Japanese government approved the operation and scheduled it for late November or early December.

The American military continued to discount intelligence that the Japanese would try such an attack. Their contingency plans for a war estimated that a hostile Japan would attack the U.S. without warning, but in the Philippines, not at Pearl Harbor. The American plan anticipated that Japan could "seize Guam and the Philippines and then go on the strategic defensive, forcing a war of attrition." This war would be one America would win, the strategists concluded, but Japan would hold the "upper hand for about a year" until U.S. material strength would take effect. The war would be "long and expensive" with "heavy losses on both sides." The U.S. Army planners estimated that victory could not occur against Japan until 1080 days after mobilization. The navy said it would take about a year longer. Reviewing all this, Marshall prompted his commanders to hasten the training of the new army forces. "We must be prepared," he wrote, "to fight immediately, that is within a few weeks, somewhere and somehow. Now that means we will have to employ the National Guard [troops] for that purpose."[16]

The ordinary soldiers of the 164th Infantry knew nothing of Marshall's memo or the war plans. But as they went about their business at Claiborne, their fates were carved into these documents.

On December 7, 1941, Melvin Bork and a couple of his buddies were enjoying a weekend pass in Baton Rouge when "a policeman came up to us and told us we had to get back to camp. We were at war." John Paulson had just returned to Claiborne from special training in North Carolina for new non-commissioned officers. He was catching some rest when another fellow came into his tent and asked him where Pearl Harbor was. Ralph Oehlke was on a target range with a real 37mm anti-tank gun that he and his crew had just received the day before. "We had waited for it for months and we had just set it up and gotten off a shot or two and this lieutenant drove up in a jeep and told us to get the gun hitched up and get back to our barracks. We might ship out that day."[17]

The next day, Colonel Sarles was informed that his regiment was now detached from the 34th Division. The regiment was ordered to San Francisco. They boarded some hastily assembled trains and left the same day. The speed of the departure fed speculations that the 164th would have to face a Japanese landing on the California coast. Among untried soldiers it is a rare man who wants his comrades to think he's afraid. So as the trains slowly headed for San Francisco, the men spoke of sports and food and girls and only occasionally of the prospect of combat. But unwelcome thoughts still floated through their minds. This was particularly true for the new noncoms. Bernie Wagner, a G Company soldier but recently advanced to squad leader, put it plainly. "When you led a group of men, their lives were in your hands; that's a lot for someone as young as we were." Another group of men were just plain angry. The regimental band had long since realized that they would indeed be expected to act as stretcher bearers and medics in time of war. They were not happy about it but accepted it. Now, as the regiment set off, a number of band members were shocked to discover that their names were listed as members of rifle companies. "Those fellows were mad as hell and I didn't blame them," Doug Campbell recalled.

Most of the trains reached San Francisco on December 12 and 13. The men were trucked to the Cow Palace for temporary quarters. There were a lot of jokes about farm boys being housed in cattle stalls. When a group of bored riflemen decided to get up a baseball game on the roof of the Palace, a rather irritated policeman stalked up to tell them to cut it out. Otherwise, the residents of the city were happy to see them. Fears of bombing by Japanese aircraft, even of an invasion, were running rampant. Doug Burtell and one of his pals in the headquarters realized how welcome they were when they sneaked out of the Cow Palace one evening and went downtown to "a fancy French restaurant where we each ordered a five dollar steak dinner. The waiter didn't let us pay, they paid for it because we were in uniform. Until Pearl Harbor, soldiers weren't very popular, people didn't want their daughters going out with us, but that changed once the war started. In San Francisco they couldn't have loved us more."

To reassure the people living on the west coast, the 164th was divided into smaller units for guard duty. Most of the First Battalion was left in San Francisco for the purpose of guarding the coastal guns that ringed the bay. Some squads were sent to guard the Golden Gate Bridge and to the harbor to keep order during a brief dockworkers strike. The men who stayed in the city were transferred to the Presidio. Life at the Presidio, a permanent army base, was pretty good. The men were amused to be awakened each morning by a cannon shot rather than a bugle call. Rudy Edwardson and his platoon "did patrols along the coast, looking for possible landings and keeping an eye on those big coastal guns." Edwardson also remembered that Company F received some new men while in San Francisco. "There was a ship that came in with some regular army troops. They had been at sea when Pearl Harbor was attacked, so they turned back. Now they were turned over to us as replacements" to fill out the rifle companies. "When we went to the South Pacific, they were unhappy soldiers to be going with us. I remember one, he had several tattoos, and he was later shot on Guadalcanal, shot through the knees. He wasn't happy to have joined our company."

Other units of the 164th left San Francisco and went to various points in the northwest for the purpose of guarding bridges and military installations. Doug Burtell volunteered to be a driver for a truck convoy heading for northern California and ultimately Oregon. "I decided that hell, that'd be more fun than the train. So we set out, me just seventeen and driving over the Oakland bridge and up the highway. We didn't have much in the way of heaters in those trucks, so it was getting pretty cold by the time we got up around Mount Shasta, where it was like fifteen below zero. We were wearing our long johns and overcoats and we were still cold. We were also in blackout conditions, trying to follow a dim tail light on the truck in front of us over those mountain roads. We did a lot of double clutching." Some of the drivers used alcohol to feel warmer. Wendell Wichmann, who commanded one of the truck convoys, was shaken when the truck he was in started to drift off the road. "Somebody had given my driver a bottle of booze at our last stop. So I had to say pull over and then do the driving while he slept in the back." Before they got to their destination, Wichmann's truck convoy stopped on Christmas day for a dinner at a small restaurant, paid for by the entire town. "They got in a choir of kids to sing for us while we ate. I've got to tell you, here were all these tough Dakota boys, and there wasn't a dry eye among us."

In the northwest, the men were sent in companies and platoons to various sites in Washington, Idaho, Montana, and Oregon. With the units spread out across the northwest, paying the men made for a challenge. Howard White had become a driver for one of the officers who distributed the men's pay. "Once a month, my duties were to carry a finance officer and the payroll to Pendleton Oregon, to Seattle, out to Idaho. We had a whole back seat [of the jeep] full of money."

Most of the headquarters company personnel were stationed at Walla Walla, Washington, alongside a grass-covered airfield. "There wasn't a Jap airplane that got near there, so we were doing a good job," laughingly remembered Elroy Greuel, a member of the headquarters company. Doug Burtell liked Walla Walla, "where everything, the movies, dances, food, everything was free. We went over to the Olympia brewery there and they'd put out cakes, pitchers of beer, all of it, and said 'Help yourselves.' Some of our guys really liked it there and ended up marrying Walla Walla girls and staying there after the war." As Gerald Sanderson and his mates arrived in The Dalles, Oregon, they expected to bunk at the city's civic center. "But there was this guy standing there and he said 'I'll take you four guys, you're going to have Christmas dinner with us.'" Sanderson asked for time to clean up after the long drive, but the man said 'Nope, you can have a bath in our bath tub.'"

The locals were ready to forgive the soldiers almost anything. Bernie Wagner, stationed at one of the railroad bridges, noted how his buddies, tiring of army rations, shot several deer out in the countryside. "A game warden came out to see us and asked if we knew anything about deer getting shot. We said, 'No, we didn't know anything about it,' but asked if he'd like to stay for dinner and have some 'beef' with us. He ate with us. As the warden left, he said 'That was pretty good beef you had, but try not to eat too much of it.'" John Paulson and his squad, guarding railroad tunnels and bridges along the Idaho-Montana border, also shot some deer and antelope for fresh meat, blandly explaining to a local sheriff that "the deer had attacked us and we fired in self-defense." After a few weeks, the men decided they liked this posting. "They could have left us out there through the whole war, we were having a good time, hunting and fishing." But they all knew this couldn't last.

Toward the end of February, all the troops were alerted to prepare to return to San Francisco where the regiment would be reassembled at Fort Ord. It also took in some more reinforcements to bring the companies up to combat strength. The new men were transferred from other units. Richard Stevens was one of these men. Stevens had joined the Kansas National Guard in 1939. But as Stevens noted in a memoir he later wrote, parts of the 35th were "orphaned" in the same manner as the 164th, and "rushed to California in response to Pearl Harbor." Then some of the men were transferred into the 164th. "Sixteen of us of various enlisted grades came to the 164th from Company M of the Kansas 137th Infantry [regiment], along with a similar number from Company M of Nebraska's 134th Infantry Regiment. We came as strangers." Stevens and his friends were "warmly received" by officers of M Company.

By the time the troops were gathered again in San Francisco, the Japanese had seized Hong Kong, much of Singapore, Malaya, Burma, and the Dutch East Indies. The American bases at Wake and Guam had fallen and the Philippines defense would collapse in April. A Japanese commercial and military empire, more than twice the size of North America, sprang up in less than four months. Japan had paid a ludicrously small price for this treasure: less than four hundred aircraft, just four warships, a few transports, and about 15,000 casualties. In Washington, some of the best American military minds were quietly estimating that it could take a decade to defeat Japan.[18]

The 164th left California on March 18, traveling on the *SS President Coolidge*, a luxury liner that was part of America's Pacific tourist fleet. The *Coolidge* had been hurriedly converted into a naval transport. Most of its staterooms had been fitted with tiers of canvas bunks, dining areas refitted into mess halls, and the ship itself camouflaged with navy-grey coloring. Still it was a luxury liner; marble columns remained in the lounges even if the chandeliers had been removed. Many of the upper deck cabins, reserved for officers, were in the words of one man "quite opulent compared to how we sailed later on." Escorted by a couple of destroyers and the cruiser

Chester, the *Coolidge* left San Francisco Bay in company with the *SS Mariposa* and the British liner *Queen Elizabeth*. Their destination was unknown. One rumor had it that they were on their way to the Philippines, to mount a last ditch defense against the Japanese. After several days at sea, they had a rendezvous with a tanker offshore from the Marquesas Islands and refueled. The destroyers departed at this point, but the *Chester* stayed with the transports which continued to head west. The liners made use of their speed – an average of about nineteen knots – to outrun any Japanese submarines. Scuttlebutt continued to be sketchy.

Many of the soldiers, landlubbers at heart, were seasick during the voyage, but others enjoyed the trip to Australia. "The food and views were fine and dandy," John Paulson remembered. Rudy Edwardson enjoyed the sea as well, but mindful of the daily submarine drills and the fact that the naval personnel would lock down all the watertight hatches if an attack developed, Edwardson and some others spent the trip on deck, sleeping near a life raft and going below only to get chow in their mess tins. No one could forget for long that when the voyage ended, their war would begin.

(Endnotes)

[1] For a good summary of the state of the American army in 1940-41 and the building of the new camps, see Geoffrey Perret, *There's a War to be Won: The United States Army in World War II* (New York: Random House, 1991), pp.3-46.

[2] William Hagen, interview with Shoptaugh, April 5, 1991, transcript, p. 16; previously cited interviews with Burtell, Paulson, and Tollefsrud; John Hagen tapes at Northwest Minnesota Historical Center (NMHC).

[3] Gerald Sanderson, interview with Shoptaugh, September 15, 2007.

[4] In addition to interview information, there are good descriptions of Claiborne in Cooper and Smith, *Citizens as Soldiers,* pp. 266-269, and in Cecil Atkinson and Kathy Tilley, *Camp Claiborne* (Forest Hill, LA: Ack Hill Publishing Co., 1990).

[5] Shoptaugh interview with William Tucker; Douglas B. Campbell memoir in the 1985 164th Regiment Band reunion booklet.

[6] Osborne Frederickson and Morris Horwitz to Dr. O.J. Hagen, March 7, 1941, enclosed with an undated letter of John Hagen to his father, both in the Hagen Family Papers, NMHC; Shoptaugh interview with William Welander.

[7] John Hagen letters of March 30, April 6 and April 20, 1941, all to Dr. Hagen, Hagen Family Papers, NMHC; Shoptaugh interview with Don Hoffman; Cooper and Smith, p. 267. Ellard Walsh, the adjutant general of Minnesota, commanded the 34th Division until August 1941. The 34th was comprised of the guard regiments from North Dakota, South Dakota, Wisconsin, and Minnesota.

[8] Shoptaugh interviews with Wiest and Hoffman; SHSND interview with Ralph Oehlke.

[9] Christopher R. Gabel, *The U.S. Army GHQ Maneuvers of 1941* (Washington DC: US Army Center of Military History, 1991), esp. pp. 184-96; Shoptaugh interview with Boyd; Harold Larson, "An Account of the 164th Infantry Regiment in the South Pacific, From Camp Claiborne to the Philippines," typescript in the 164th Infantry Association Papers, UND. Cooper and Smith, *Citizens as Soldiers,* p. 269, summarize the officer changes as follows: "Through no fault of their own, many National Guard officers were overage for their rank and undertrained for their role in combat. . . less than forty percent of all National Guard officers had had the opportunity to complete a course in one of the [regular army] service schools." As a result, guard officers frequently "fell short of the demands of combat command" and "eventually received orders assigning them to less rigorous duties." True enough, but the guard soldiers still resented many of the changes.

[10] Wendell Wichmann, interview with Shoptaugh, September 14, 2007.

[11] Shoptaugh interviews with Welander and Ferk.

[12] The incident with Sarle's staff car is related in "Lessons on Borrowing a Staff Car," *The 164ᵗ Infantry News,* December 2000. Edwin Kjelstrom memory in 164ᵗʰ Regiment Band reunion booklet for 1983. See also Cooper and Smith, p. 268.

[13] John Hagen to O.J. Hagen, July 16, 1941, both in Hagen Family Papers, NMHC; interviews with William Hagen, Paulson, Tollefsrud, and Welander noted the bitterness among men over the draft extension. The Railey report is summarized in Lee Kennett, *GI: The American Soldier in World War II* (New York: Scribner's, 1987), pp. 69-71.

[14] Cooper and Smith, pp. 267-68; Shoptaugh interviews with Burtell and Tollefsrud; Robert Sanders memoir in the 164ᵗʰ Regiment Band reunion booklet for 1985.

[15] This story is taken from a 2007 interview with James Fenelon and Les Aldrich, conducted by Merry Helm of Fargo, who kindly shared a copy with me.

[16] Agawa, *The Reluctant Admiral,* pp. 228-29 for Japanese war games; Henry G. Gole, *The Road to Rainbow: Army Planning for Global War, 1934-1940* (Annapolis, MD: Naval Institute Press, 2003), pp. 84-86, 98-100. See also Edward S. Miller, *War Plan Orange: The U.S. Strategy to Defeat Japan* (Annapolis, MD: Naval Institute Press, 1991), pp. 250-66.

[17] This entire section is based on Shoptaugh's interviews with Bork, Burtell, Edwardson, Sanderson, and Wendell Wichmann; Shoptaugh interview with Elroy Greuel, July 24, 2007; SHSND interview of Oehlke and Wagner; the Campbell memoir in the 1985 164ᵗʰ Regiment Band reunion booklet; Howard White interview by James Fenelon; Richard Stevens, telephone interview with Shoptaugh, August 29, 2007.

[18] H. P. Willmott, *Empires in the Balance: Japanese and Allied Pacific Strategies to April 1942* (Annapolis MD: Naval Institute Press, 1982), is an excellent summary of the early Pacific campaign.

Chapter 4: New Caledonia, "A Lazy Man's Paradise" South Pacific: 1942

Les Wichmann wanted to see land again, any land. Wichmann was no sailor. Like so many of his friends, he had not left North Dakota much before going to Louisiana in 1941, and the ocean voyage from San Francisco was not what he had expected. The *Coolidge's* voyage across the Pacific had been a queasy ordeal, for him and many others. The liner had pitched heavily in a storm one night, which had driven him to find a more secure bunk. "I went to the main lounge, and there were sofas there, back to back, bolted down. I got into the space between two of those sofas, but the ship's rolling kept me sick all night."[1]

So Wichmann was relieved when the *Coolidge* made port in Australia on April 7. Other ships in the convoy went on to Sydney, where elements of the 41st Infantry Division landed before going on to New Guinea. The 164th offloaded in Melbourne. The Australians were facing a possible Japanese invasion and were happy to see the GIs. "They went wild over us," Wichmann remembered, "waved and were certainly happy to see us. As we marched through part of the city, the people gave us candy and cookies." The regiment spent about four days in Melbourne. Most of the men were kept busy transferring equipment and supplies from the liners to three smaller ships. "We did twelve hour shifts moving gear into these three Dutch ships," Dick Stevens recalled. Some others got a chance to see parts of Melbourne. Doug Burtell and Bob Dodd borrowed a few dollars from one of the lieutenants and took off to see part of the city. Philip Engstrom, a member of Company F, had been seasick the whole way across the Pacific and was delighted to be on dry land, even if the inhabitants "drive on the wrong side of the road here." He promised himself that if he got back home he would "never take another trip on the water – no siree!"[2]

Dennis Ferk took a street car out to the city outskirts and back. John Hagen visited the Melbourne zoo and thought it was "one of the finest in the commonwealth." But the unit had to resume its journey, and a few of these tourists had to rush back to the docks and board the ships at last minute. They set sail in three Dutch steamers, the *Maatsuyker,* the *Kremer* and the *Van Huetz,* tramps that had plied the waters of the South Pacific for years. It would be a memorable trip.[3]

The steamers were bound for New Caledonia, the central base for the American campaign that was developing for seizing control of the Solomon Islands. The harbor at Noumea, in New Caledonia, was too shallow to accommodate a ship the size of the *Coolidge,* so the Dutch ships had been used. Dick Stevens remembered that some machine guns had been "lashed to the rails," and two anti-tank guns had been set on the decks of the *Maatsuyker.* Stevens did not find the ship so bad, but other men cursed every minute of their stay aboard the smaller craft. Wichmann travelled on the *Maatsuyker.* Still no lover of the sea he found the ship's rails were so coated in rust he was afraid they might give way in he leaned on them. He and his comrades thought it "was just a stinking cattle boat." They feared that a Japanese sub might come along and because the "hull wasn't more than a half-inch thick, one torpedo could have sunk the thing so fast it would have been a total loss of passengers and crew."

Like Wichmann, Leo Swanson, a member of the band with additional duties in the Headquarters Company, was certain that if anything went wrong these ships would sink like stones. The lowest deck of his steamer, the *Kremer,* leaked; "once we were out in the ocean, water kept running over the floor [i.e. deck] so our barracks bags had to be hung up on our bunks. The rats kept running over the partitions that held our bunks, which were tiered." Only the officers of the ships were Dutch, and they kept to themselves on the upper decks, which Swanson said "were barred off with iron bars . . . It made you think of a ship that was used for prisoners or perhaps an old slave ship." The crews on these tubs were mostly Javanese, who subsisted mostly on a diet of potatoes and dried fish. On the *Maatsuyker* the fish were piled on the afterdeck, covered by a stiff, salt encrusted tarpaulin. When the soldiers learned that they were expected to eat this fish they were appalled. "Even the medical officers said the food was unfit to eat," Swanson noted. "The Javanese would sit with their feet in the tub with the peeled potatoes and water." The GIs lived mainly on bread that some of the cooks managed to steal from the officers' galley. Swanson admitted that because of "the bread that was stolen and some of the officers' beer that happened to disappear, there got to be very hard feelings between the officers and the enlisted men."[4]

On April 19, the tramps sailed into Noumea harbor. Noumea was the capital and largest town in New Caledonia, a French colony located southeast of the Solomon Islands chain. The men were glad to be on land again. It took days before they fully recovered their "land legs." American troops had been on New Caledonia since the first week in March. In late January 1942, a hastily assembled scratch force, made up of two infantry regiments with supporting medical, artillery and engineering units, had departed New York in seven transports, bound for the Pacific with the designation of Task Force 6814. These two infantry regiments, the historian of the Americal Division later wrote, "were those declared surplus when the 26th and 33rd Infantry Divisions, recently federalized Massachusetts and Illinois National Guard Divisions, respectively, were reorganized." The Task Force, he noted, "might have looked a bit like an infantry division but, for the most part, it appeared as nothing but an odd conglomeration of spare parts, a wartime military stew of men and equipment."

Task Force 6814 was sent to New Caledonia to help stem the Japanese tide that was washing over the Pacific. The arriving GIs knew almost nothing about the island. Joe Marseglia, a member of one of the field artillery units attached to the Task Force, remembered that the men were told to "always carry your rifle with a round in the chamber" because "there were headhunters" on New Caledonia. He was surprised to find on arrival that the island population was quite civilized. Japan had New Caledonia on its list of eventual conquests. Had they been able to seize it, the island would have been of inestimable value. New Caledonia was a large island and was also rich in resources, including copper, gold, manganese, cobalt and especially nickel deposits, which Japan could put to good use in military production. But the real prize was New Caledonia's strategic position, resting astride the routes from the Fiji-Samoa region to Australia or New Zealand. With New Caledonia, Japan could isolate Australia and threaten its east coast ports. Conversely, from New Caledonia, American forces could mount a drive up the Solomons chain and neutralize the Japanese bastion at Rabaul on the island of New Britain. New Caledonia was thus a major linchpin for the American Pacific strategy.[5]

The same week that Task Force 6814 arrived on New Caledonia, President Roosevelt promised British Prime Minister Churchill that the United States would assume responsibility for defending Australia and New Zealand. But in March of 1942, the American ability to hang on to New Caledonia was in real doubt. Even when reinforced by the 164th, the makeshift division lacked the men to defend every possible beachhead where Japanese troops could land. All the Allies had on hand for naval support were a few American and Australian cruisers and destroyers and a handful of aircraft. This was not much to pit against the full might of Japan's fleet.

One thing that the Allies did have on their side was an unparalleled look over the shoulders of the enemy. U.S. military services had made enormous strides in intelligence gathering and in cryptology. A code-breaking project, itself disguised under the rubric "Magic," had penetrated Japanese diplomatic codes before Pearl Harbor. The Americans had now cracked the primary Japanese naval code, giving them insights into the enemy's plans. In April, the commander of the Navy's code breakers informed his superiors that the Japanese now were planning two major operations. The first, he stated, was designed to isolate Australia by seizing bases on New Guinea and moving down the Solomons toward New Caledonia. The target of the second operation was not yet clear, but the intelligence gathered did indicate that it would be a very large operation and would involve the most powerful parts of the Japanese fleet. This latter operation would lead to the crucial Battle of Midway.[6]

The threat to New Caledonia was made clear to the 164th from the moment it arrived. The regiment was ordered to set up a tent camp a few miles north of Noumea. Soon after, some men were sent out to act as radio operators on an Australian patrol boat that was looking for Japanese submarines. Howard Lauter, a soldier in H Company, recalled that his unit was trained to use their machine guns and mortars in "a full-scale beach defense using live ammo." Lauter was impressed by the coordinated firing of the 155mm guns that had been dug in to repel a possible landing. The elements of Task Force 6814 and the 164th were formally united as an infantry division. The division's commander was Major General Alexander M. ('Sandy') Patch, who had been personally selected by George Marshall to take charge at this critical spot in the Pacific. Perhaps hoping to create some sense of *esprit de corps,* Patch asked the men to suggest names for this new division. The winning choice was submitted by a private in the 26th Signal Company who suggested "Americal," which he had cobbled together from **Ameri**can troops in New **Cal**edonia.[7]

The Japanese aside, New Caledonia was a pleasant place to be stationed. Dr. Arthur King, the head of Task Force 6814 medical staff, wrote a "sanitary report" for the island in which he commented on the conditions on

New Caledonia. The water supply was ample and safe, the food supply good and the sanitation, while primitive in places, was decent. Most important, King noted, was the fact that the anopheles mosquito was not present on New Caledonia, so that malaria, the scourge of other islands, was not a danger to the men. Nor was there great danger from many of the other fevers found in the Pacific rain forest islands: dengue, yellow, or typhoid. He did warn that the GIs could suffer from intestinal problems, such as amoebic dysentery, unless food was carefully inspected, and fresh fruit and vegetables were obtained from Australia. With these precautions, he concluded that "health in general" should be "excellent."[8]

King was ordered to take care of another task in connection to the men's health, one that he was ordered to keep strictly secret. As he explains in his postwar memoir, he was told to quietly set up a brothel on New Caledonia. The French, concerned about the safety of the women on the island, had advised Patch to establish a "military bordello." Patch gave the job to King, who worked with the local Provost Marshal to find a large house "surrounded by a large wall, to control access." A "former madam" was persuaded to run the operation and a retired policeman agreed to "maintain order" in the house. MPs guarded the entrance, while a side exit led directly to a "prophylaxis tent" staffed by an army doctor. Once the house was in operation, Patch sent word to King to remember that the "divisional commander" was "never to know that the house exists."

The men of the 164th spent about six months on New Caledonia and generally remember it as a pleasant interlude in an otherwise grim war. A hurriedly prepared army guidebook given to the new arrivals stressed that the Americans had "come to New Caledonia . . . as friends to help guard [this] strategic interest of the Allies in a distant and vital corner of the world." The men were ordered to maintain decent relations with the inhabitants. John Hagen and Bill Boyd certainly tried to do that. They purchased horses from a local farmer and rode in the hills outside Noumea. Hagen wrote his father to send him more movie film so he could record the beautiful "sunsets over the mountains." He thought New Caledonia was a "lazy man's paradise." Men attached to the headquarters and supply units kept peacetime hours, working eight hours before knocking off for the day.[9]

The riflemen were busier. Howard Lauter and his buddies "spent many days learning how to cross rivers by ropes, how to climb on cargo netting, practice hand to hand combat, and how to throw hand grenades." PFC Phil Engstrom thought most training "real old Claiborne style again," a routine of marches, inspections and lectures that was mostly "Boy Scouts" stuff. He and his pals relieved the tedium of Army food by trading some of their worn out fatigues to a French family for chickens. Doug Burtell, Bob Dodd, Ed Mulligan, and others went on an expedition to cross the island from north to south. Burtell thought the purpose of the trip was to test how readily a Japanese unit could cross the island if they invaded. Whatever the intention, the men learned how to pack for long-range marches. "They gave us large cans of rations for like 20 guys, but they were too big to carry, so we broke those down and took only the light items – candy, chocolate, oatmeal, some dried fruit. We were running out of food near the end." Fred Maier, another member of the trek, was very hungry by the time they ran across some wild pigs. "It was just like the OK Corral the way we were shooting. Lucky we didn't kill each other, but we finally got one of the pigs." A French engineer who lived at the mines also gave the men some food.[10]

While the men trained, the course of the Pacific war changed dramatically. In May, American aircraft carriers blocked a Japanese attempt to seize Port Moresby at the Battle of the Coral Sea. This was a signal victory for the defense of Australia and it opened the way for General Douglas MacArthur to begin a long campaign against enemy bases in New Guinea (see Map 1). A few weeks later, the Americans surprised the Japanese fleet off Midway and sank four enemy carriers. These actions, made possible by the broken enemy code signals, turned the war around for the Allies. The Japanese lost not only the bulk of their naval air power but hundreds of their best airmen, the key to their early victories. The Americans lost a carrier of their own and a few other ships. But the Japanese retreated and Midway and the American victory made it possible for the U.S. Navy to mount an offensive in the South Pacific.[11]

Events moved quickly after Midway. On June 8, MacArthur submitted a plan to Washington for an assault on Rabaul. All he needed to supplement his forces already at hand, MacArthur said, was a task force with two carriers and a Marine assault division. The Navy however did not wish to put any of their carriers under MacArthur's command. Admiral King, head of the Navy, argued that the force available was too small to smash the Rabaul bastion. He offered up his own plan – to use New

Caledonia as a base of operations for seizing positions in the southern Solomon Islands, which he listed as the island of "Tulagi and adjacent positions." After that, he concluded, the Americans should work their way toward Rabaul by seizing other bases in the Solomons. Rather than wait for this suggestion to be discussed and approved, King ordered Admiral Chester Nimitz, his overall commander in Hawaii, to prepare a detailed plan for these operations.[12]

Both MacArthur and King were engaging in a high-stakes poker game. MacArthur wanted full command of the Pacific War, and went so far as to tell Marshall that he would resign his commission rather than take orders from the Navy. King was equally adamant in insisting that the Navy must dictate strategy in the Pacific. But behind the scenes both men knew that Roosevelt and General Marshall had agreed with the British that the war in Europe must take first priority, and that the seven American divisions in the Pacific were expected to stand on the defensive. So in order to get American troops engaged in an offensive against the Japanese, King and MacArthur could work together, at least for the moment, to convince Marshall to let them launch two "limited offensives." MacArthur would use American and Australian forces to blunt the Japanese advance in New Guinea, and King would use the First Marine Division to seize Tulagi. The seizure of Rabaul remained the ultimate goal. This, they argued, would end the threat against Australia, drive the Japanese back toward the Philippines and open up further possibilities in Pacific. Marshall was persuaded enough to convince Roosevelt to authorize two Pacific commands, MacArthur's Southwest Pacific

Area, and the Navy's Pacific Ocean Area. MacArthur and King could begin their "limited attacks."[13]

A few weeks later, American reconnaissance planes discovered evidence that the Japanese were laying out an airfield on the island of Guadalcanal, across the sound from Tulagi. The Navy advanced its date for invading Tulagi to early August, and added Guadalcanal to the objectives. Trying to seize and hold both Tulagi and Guadalcanal with the limited forces available at the time was a bold decision. Some thought it verged on foolhardy. Vice Admiral Robert Ghormley, commander of the South Pacific theater, studied the plan at his headquarters in New Caledonia and expressed grave doubts that the it could succeed. He thought the Allies lacked enough aircraft or ships to adequatcly support two offensives, and that Japanese aircraft at Rabaul would pose a grave threat to Allied ships in the Solomons, especially "during the approach, the landing, and the unloading" of the American forces. But King ordered the attack to proceed.[14]

And so began the chain of events that would put the 164th regiment on the shores of Guadalcanal. The campaign opened on August 7 when units of the reinforced First Marine Division stormed ashore on Tulagi and Guadalcanal. The enemy on both islands were completely surprised; those on Guadalcanal, labor troops mostly, fled into the jungle. The marines seized the half-finished airfield together with quantities of food and supplies. The marines however quickly learned that their adversary was formidable when the Tulagi garrison fought to the last man.[15]

For the moment, the invasion appeared to be an unqualified success. But then errors were made that turned the campaign into a protracted slugfest. Some of the errors stemmed from the way in which command in the region was divided between MacArthur in Australia and Ghormley in New Caledonia. For example, reconnaissance photographs taken of Guadalcanal in July, which would have been priceless to Vandegrift and his planners, were misplaced by MacArthur's command in Australia and would not turn up until six months later. Another problem emerged when Ghormley, who doubted the success of the campaign, acted to protect the Navy's ships even when such action exposed the marines to greater danger. Major General Alexander A. Vandegrift, First Marine Division commander, had stipulated that he needed the supporting naval forces to remain on hand until his marines were well entrenched, at least five days. But when Admiral Frank Jack Fletcher, who commanded the three carriers that made up the American air cover, decided that he would stay in range of Japanese aircraft for no more than two days, Ghormley took no action later to overrule Fletcher. Fletcher withdrew his carriers on August 9, leaving Vandegrift with no air cover.[16]

The Japanese at Rabaul reacted vigorously to the American attack. They launched air strikes on both August 7 and 8, and sent to sea an ad hoc surface force consisting of seven cruisers and one destroyer. It would arrive off Guadalcanal in the deep dark on August 8-9. An amphibious patrol plane spotted this strike force but a communications failure prevented the warning signal from reaching Guadalcanal until it was too late. The American cruisers and destroyers off Guadalcanal, still in the dark about the approaching enemy, were surprised by the Japanese cruisers around midnight and badly mauled. Four American and Australian cruisers went to the bottom, as did a destroyer. Three other warships were badly damaged. Over a thousand sailors died. The Japanese sailed away with scarcely a scratch. The failure of the Japanese admiral to press on and destroy the American transports saved the landing from a complete disaster. As it was, the American transports and wounded warships fled back to New Caledonia, leaving behind a handful of landing craft and 25,000 angry marines on Guadalcanal and Tulagi who had no air cover and not enough supplies to properly defend their positions.[17]

The Japanese were left in control of the waters around Guadalcanal and could reinforce their surviving troops on Guadalcanal almost at will. Guadalcanal had seemed like a quick victory to the Americans just twenty-four hours before. Now the survival of Vandegrift's force was in doubt. But the Americans had possession of the airfield. With the help of Seabees who rushed to the island in destroyer transports and amphibious aircraft, the marines were able to get the airfield into good enough shape to permit the 1st Marine Aircraft Wing to fly in fighters and dive bombers to help preserve the perimeter. The Japanese meanwhile used destroyers and transports on fast nightly runs to bring reinforcements to the island.

Throughout August and September, the battle for control of Guadalcanal raged back and forth. The marines drove back every enemy attack to capture Henderson Field. The fanaticism of the Japanese in these attacks – the way they were willing to expend the lives of their men for a few yards of ground – shocked even the most hardened marines. But the American Navy lost several vital ships, including an aircraft carrier in keeping Vandegrift's men supplied. The American ship commanders hesitated to remain in the waters around Guadalcanal for more than a few hours, which only increased the marines' feelings of isolation. The American perimeter held, but it was so small that there was no "rear area" where a man could feel safe. Every man understood that his life was hanging in the balance all the time, that somewhere nearby, the Japanese lurked, be it only by a sniper looking for a target or a mortar shell being ranged onto the airfield. The Japanese bombed the Americans at least twice each day and once at night. Japanese destroyers frequently shelled the lines after delivering another contingent of enemy reinforcements.

Admiral Nimitz sent a special communiqué to spur his naval commanders to greater efforts: "We cannot expect to inflict heavy losses on the enemy without ourselves accepting the risk of punishment." What was needed most was more aircraft at Henderson Field to protect the American lodgment and more ground troops to reinforce the marines. The spread of malaria was reducing their ranks more then the Japanese attacks. When General Millard Harmon, the overall commander of U.S. Army units in the South Pacific, warned Ghormley on October 6 that Vandegrift had to be "materially strengthened" or Guadalcanal would be lost, the decision was made to rush an Army regiment to the island. Ordered by Ghormley to select one of his regiments to send to Guadalcanal, Patch chose the 164th Infantry.[18]

The 164th had kept busy while the Guadalcanal fighting continued. "Training activities" and "usual staff meeting" are recorded again and again in the regimental log for August and September. When a trawler loaded with artillery ammunition sank in shallow water off Noumea, work crews from the regiment joined with navy divers to salvage as many of the shells as possible for shipment to the marines. The men were aware that they would probably be sent into the combat zone very soon. Two marine officers, recovering from wounds received on Guadalcanal, visited the companies and talked to the men about "fighting the Japs." Both marines emphasized that "the Jap soldier will never surrender, you have to kill him." They also warned the soldiers to beware of Japanese deceptions. Alvin Tollefsrud was now a squad sergeant, and he remembered "Major Bailey, he was a gung ho guy. He had been wounded and came to New Caledonia for treatment. He said the Japs would make noise like they were a wounded [American] man. But don't go out to them because they'll kill you. He went back and won the Medal of Honor, but was killed by a sniper."[19]

On September 14 a very significant change occurred when Earle Sarles was relieved from command of the regiment. "Col. Sarles relieved due to over age," says a penciled entry in the Regimental log. Patch had given the order, which is interesting because Sarles was then 55 years old and Patch himself was 53. There are no indications that Sarles was experiencing any illness or health problems. "It was hard to see Colonel Sarles leave the regiment as he had been with the regiment since 1905 and was a father to us all," Sam Baglien wrote in his diary. Sarles was sent back to the states.[20]

The new commander was Colonel Bryant Moore, a West Point graduate of 1917, part of the Task Force 6814 expedition and until September the American's assistant chief of staff for intelligence. He had attended both the Infantry School at Fort Benning and the Staff College at Fort Leavenworth, the traditional pathways for combat command. When Moore arrived at the headquarters of the regiment, persons there recalled that he announced, "I'm Bryant E. Moore and I'm here to relieve Colonel Sarles." Those at the HQ who had liked Sarles and his more informal ways took offense at Moore's abrupt manner. "We just didn't like Moore, he wanted to throw his weight around," Jim Fenelon later recalled. Several officers, including apparently all three battalion commanders, were dismayed that the new commander was not one of their own. Sam Baglien, the regimental exec, was regarded by several company commanders as their choice for a new leader.

Officers with long Guard experience and ties to the "guard community," were fondly regarded by many of the men. But while such familiarity helped morale, would it keep men alive in the heat of combat? The regular Army's priority was to select an officer that it regarded as fully trained to command in a combat zone. Moore was the man that the Army chose, and it was inevitable that some of the Guard officers took offense at this. Baglien soon began to criticize every action by Moore, not only in his diary but to other officers.[21]

When the regiment was ordered to Guadalcanal two weeks after Moore took command, a rumor flew through the ranks that the 164th was chosen as the "most expendable" unit. "North Dakota didn't have that many people, not like Massachusetts [home of the 182nd regiment] or Illinois [home of the 132nd]," one veteran of the regiment stated fervently in 2007. "So if we were lost the uproar would be a lot less." But Patch, who would have hurt his own career in deliberately selecting a unit he expected to be defeated in battle, most likely picked the 164th because it was stationed closer to Noumea than either the 132nd or 182nd, both encamped in the northern part of New Caledonia. Speed was of the essence.[22]

Boarding the *McCawley* and the *Zeilin* on October 9, the men of the regiment set off. The two transports carried the soldiers, over 200 men to reinforce the First Marine Air Wing and 85 marine riflemen as infantry replacements. As with the previous voyages, several of the GIs were seasick. They were also uncommonly quiet. "We gave up our old World War I style helmets when we left Noumea," Doug Burtell remembered. "We got the new helmets just as we left. We knew we were going into combat then."

As already recounted, the Japanese welcomed them to the green hell of Guadalcanal on October 13 with a storm of bombs and shells. "I, along with a few thousand other boys, admit that we were scared beyond all limits," Phil Engstrom wrote in his diary. More men died from the shelling over the next few days, including the long-serving and popular Captain George ("Jug") Newgard, who was hit by a dud shell. Newgard had most of his legs torn off, and would have bled to death immediately but for the efforts of three men – Russell Opat, Jack Simmons, and Barney Deering – who carried him to the hospital. But he couldn't be saved. "Jug died this morning," Baglien recorded on the 16th. "He is buried in the Marine cemetery. I have written Pat [Newgard's wife]."

Everywhere the soldiers went the marines told them that a major Japanese infantry attack was expected at any time. "This is a peculiar war," Sam Baglien wrote in his diary on the 19th. "We have an airport [but] the Jap Navy hits us in the ass and we fight them to our front. They bomb the shit out of us from the air. Ho hum. That's something for the Masters to ponder over."[23]

No one would have long to ponder. The Japanese were already on the move in the jungle south of the American perimeter. Their assault was just days away.

(Endnotes)

[1] Les Wichmann described his Pacific experiences in a tape he recorded for Robert Dodd, Jr., who kindly provided me a copy.

[2] Richard Stevens, telephone interview with Shoptaugh; Robert Dodd Jr., *Once a Soldier*, p. 40; The diary of Philip Engstrom (which his widow kindly allowed me to copy), entry for April 8, 1942. April 7 in Australia was April 6 in the U.S., due to the International Date Line.

[3] Shoptaugh interview with Ferk. John Hagen also noted in his April 19, 1942 letter to his father that he could not mention which "metropolis in Australia" the unit had briefly visited because "the censors don't appreciate it." A side note: two members of the 164th Regiment deserted while in Melbourne, by hiding out in the bowels of the *Coolidge*. They were not discovered until the ship was close to the U.S. Both were court-martialed and imprisoned. One later returned to North Dakota, where one veteran of the unit said, "we'd see him from time to time, but few of us would ever speak to him."

[4] Stevens telephone interview; Les Wichmann tapes; Leo Swanson memoir in the 1983 164th Regiment Band reunion booklet.

[5] Cronin, *Under the Southern Cross*, pp. 4-6; Gordon L. Rottman, *World War II Pacific Island Guide* (Westport, CT: Greenwood Press, 2002), pp. 66-71. Joe Marseglia's memory of being warned about headhunters can be found in William McLaughlin, *The Americal Generation* (Peaks Island, ME: Cape Elizabeth Press, 1999), pp. 17-18. McLaughlin's memoir is a very good account about the Massachusetts troops in the Americal Division, especially the artillery components.

[6] Ronald Lewin, *The American Magic* (New York: Penguin Books, 1982), pp. 87-91; Kenneth Macksey, *The Searchers: Radio Intercept in Two World Wars* (London: Cassell, 2004).

[7] Howard Lauter information from an essay Lauter wrote before his death. Frank Albrecht, who knew Lauter in Pennsylvania, kindly provided a copy of this essay. For the Americal, see, Cronin, pp. 28-29, and David Taylor, "In the Beginning – Ameri – Cal" in the *Americal Newsletter,* April-June, 2006. The US Army formally designated the Americal as the 23rd Division.

[8] Arthur King, "Sanitary Survey of New Caledonia," May 21, 1942, Arthur King Papers, American Jewish Archives, Jacob Rader Marcus Center, Cincinnati, Ohio. King's sanitary report was used in the making of the *Pocket Guide to New Caledonia* (Army Service Forces, 1942), a joint Army-Navy publication that several 164th men obtained and kept. Douglas Burtell provided a copy of this guide to the author.

[9] "Pink house" described in Arthur King, *Vignettes of the South Pacific* (Cincinnati: Privately published, 1991), pp. 31, 59-61. Island conditions in

Pocket Guide to New Caledonia (Army Service Forces, 1942). See also John Hagen tapes, and Hagen to Olaf Hagen, July 2, 1942, Hagen Family Papers.

[10] Lauter's memoir; Engstrom diary, entry for June 28, 1942; Burtell, Dodd, Mulligan, and Fred Maier, taped memoirs, copies courtesy of Robert Dodd, Jr.

[11] The finest works on Midway are Gordon Prange *Miracle at Midway* (New York: McGraw Hill, 1982) and Jonathan Parshall and Anthony Tully, *Shattered Sword: The Untold Story of the Battle of Midway* (Washington DC: Potomac Books, 2005). For additional detail see the documents in Frederick D. Parker, *A Priceless Advantage: U.S. Navy Communications Intelligence and the Battles of Coral Sea, Midway, and the Aleutians*, a documents microfilm issued by the National Security Agency, Center for Cryptologic History, 1993. The Yamamoto quote is in *The Reluctant Admiral,* p. 310.

[12] Thomas B. Buell, *Master of Sea Power: A Biography of Fleet Admiral Ernest J. King* (Boston: Little, Brown and Co., 1980) pp. 214-217. See also H. P. Willmott, *The War with Japan: The Period of Balance* (Wilmington, DE: SR Books, 2002) which lays out the full strategic options for each side at this point in the war.

[13] Several military historians, including Ronald H. Spector, *Eagle Against the Sun: The American War with Japan* (New York: Free Press, 1984) and Harry Gailey, *War in the Pacific: From Pearl Harbor to Tokyo Bay* (Novato, CA: Presidio Press, 1997) have argued cogently that by dividing the Pacific war into two commands under MacArthur and Nimitz, the American government delayed victory. But it is at least arguable that it took the pressure of both MacArthur and King to get any Pacific offensive underway in 1942.

[14] Detailed accounts of the origins of the New Guinea and Guadalcanal campaigns can be found in John Miller, *Guadalcanal: The First Offensive* (Washington DC: Department of the Army, 1949); Frank O. Hough, *Pearl Harbor to Guadalcanal: History of U. S. Marine Corps Operations in World War II* (Washington DC: Government Printing Office, 1958), and Richard B. Frank, *Guadalcanal* (New York: Penguin Books, 1992). Two important books on the influence of European strategy in early Pacific planning are Rick Atkinson, *An Army at Dawn* (New York: Holt, 2007) and Richard W. Steele, *The First Offensive, 1942* (Blooming: Indiana University Press, 1973).

[15] One of the first casualties in the long Guadalcanal campaign was Adrien Vallancourt, the cousin of Frank Albrecht, who provided this story. Vallancourt was injured and five others killed on August 7 when a powder bag exploded aboard the *USS San Juan*. The Navy lost many more men in the Solomons campaign than the Army or Marines, often because of mistakes made by Captains who underestimated the Japanese navy's skills. On this, see Eric Hammel, *Guadalcanal: Decision at Sea* (New York: Crown Publishing, 1988).

[16] Ghormley did not attend the final briefings for the Guadalcanal landings but sent a representative from his staff to them and had to be aware of Fletcher's

intention to withdraw his carriers. Ghormley thus either failed to realize how vulnerable an amphibious force would be in the early days after landing or decided the carriers were a greater priority than the First Marine Division – either way, his inaction here and later in the campaign ultimately cost him his South Pacific command.

[17] Here and below, consult Vandegrift, *Once a Marine* and Frank. See also the "Navy Seabees on Guadalcanal," by Larry G. DeVries, at http://www.seabeecook.com/history/canal/cactus.htm, and Huie, *Can Do*. For details of the naval battles around Guadalcanal consult Samuel Elliot Morison, *The Struggle for Guadalcanal* (Boston: Little Brown, 1949), and Eric Hammel's two books, *Guadalcanal: The Carrier Battles* (New York: Crown Publishing, 1987) and *Guadalcanal: Decision at Sea*.

[18] How the reinforcement of Vandegrift was determined is discussed in detail in E.B. Potter, *Nimitz* (Annapolis, MD: US Naval Institute Press, 1976), pp. 191-95, from which quotes are taken, and in Frank, pp. 276-77. Edwin Hoyt, *How They Won the War in the Pacific* (New York: Weybright and Talley, 1970), pp. 151-158, also provides a detailed account and quotes from the official record of Nimitz's meetings with Ghormley and others on New Caledonia. This record states that Ghormley claimed that Patch's division had "only one regiment [that] was trained for combat" but the exact regiment is not identified. Patch did not attend this meeting.

[19] S2 Journal entries for August and September, 1942, 164th Infantry Records. The memory of being lectured by the marine officers is recounted in several interviews, and the quote is from a conversation with James Fenelon of Co. F. The Bailey that Tollefsrud referred to in his interview with Shoptaugh is Kenneth D. Bailey, whose actions and death are recounted in George W. Smith, *Do Or Die Men* (New York: Pocket Books, 2003).

[20] S1 Journal entry for September 14, 1942; Baglien diary entry for September 15, 1942.

[21] Moore's taking command has remained a controversial issue with original members of the 164th regiment. James Fenelon, as his comment suggests, deeply disliked him. Douglas Burtell, Melvin Bork, and James Beaton, all attached to the regimental headquarters, related stories of Moore's arrogance and insensitivity to the men. But others defended Moore, as will be seen in later chapters.

[22] The remarks are from Alvin Tollefsrud, to Shoptaugh, August 28, 2007.

[23] Burtell interview with Shoptaugh; Philip Engstrom diary entry for October 22, 1942; Baglien's original diary, entries for October 16 and 19, 1942. Opat, Simmons, and Deering information is from clippings in the Russell Opat Papers, copies given to Shoptaugh by his family. All three men were awarded Silver Stars for carrying Newgard to the hospital while the Japanese were still shelling the positions.

Chapter 5: "These Farm Boys Can Fight" Guadalcanal: October 1942

Philip Engstrom wrote in his diary for October 22, 1942: "Sooo much has happened I've missed [writing] about 8 or 9 days in here. '2nd battalion to the front lines' – [the command] came the second day we were here and up we all went, Co. F right with 'em. We have been on the line a wk. now, no actual combat with the Japs as yet . . . I'm supposed to destroy this diary, but I'm hanging on to it as long as I can . . . . I've seen boats blow up, bombs fall & hit full fuel tanks, planes falling & burning, shells whizzing overhead, men in a panicked stampede. What a terrible thing war is."[1]

The men considered entering the line to be a kind of relief after two days of unloading and stacking supplies. During the unloading Colonel Moore intervened in an argument on the beach. Moore blamed a company commander for a snarl in the unloading and ordered him to exchange jobs with his company sergeant. Moore, a marine observer thought, took charge "like a czar," which some of the older Guard officers resented. The men were more annoyed with the marines referring to them as a "stevedore's regiment" and wanted to prove themselves as warriors.[2]

On October 15, the regiment's Second Battalion marched south of Henderson Field to replace marines on the east side and southeast corner of the perimeter. Arthur Timboe, the battalion commander, watched his men move into the line with great pride. He had trained for years for just such a moment as this. He had just received word that his wife had given birth to a daughter,

Mary. Sergeant John Stannard led Company E's Third Platoon up to the point where the line turned westward. This position placed them in the front line but further away from the airfield, which Japanese planes and guns continued to attack. Like the marines, the soldiers by now were referring to the Japanese heavy artillery as "Pistol Pete."[3]

Once the men had arrived at their place in the line, Stannard's platoon began to file into the positions vacated by tired marines. The position was made up of "two-man bunkers" on a rise in the ground, Stannard wrote later, "complete with firing apertures and overhead cover of logs, sandbags and loose dirt." The sergeant was pleased to see that captured bags of rice had been placed over the top of each bunker, the rice sprouts providing the men with "some green camouflage." Stannard wondered how his men would behave once the Japanese attacked. One man that he would not have to worry about was Private Lester Kerbaugh. A native of Williston in the far northwest of North Dakota, Kerbaugh was an interesting example of how the Guard sometimes shaped a man's life. Kerbaugh had been in and out of minor scrapes as a kid, and one day he did something that brought him before a local judge. Your choice, the judge told him, join the Guard or go to jail. Kerbaugh joined Company E and started to grow up. Once at Guadalcanal he would earn a reputation for coolness under fire. A future general who served with him said of Kerbaugh: "He was skilled in all aspects of small arms combat. There was a feeling that if you were with him, then you would be okay."[4]

Coming up to the perimeter with Stannard's platoon was Lieutenant Charles Walker and a machine gun crew from Company H. "Our Marine guide led us into the bush and along a narrow trail to our Second Battalion position," Walker wrote in his own memoir. "The Marines we were relieving greeted us with enthusiasm, for they were eager to get off the defensive line for rest. As they loaded onto weapons carriers, they were all smiles." Walker carefully placed his .30 machineguns along the line to cover the ground beyond the barbed wire. The young lieutenant felt fortunate to have made it to Guadalcanal. Just a few days before the regiment set sail for the island, he was in the hospital on New Caledonia. A doctor had diagnosed him as having a "rapid heartbeat," and had cut orders for him to go back to the states. Walker was certain that he was fine, and believed that he was really being punished for "insubordination." He had argued with his commander several times and once, when the commander, "a heavy drinker," had taken a swing at him, had knocked him down. "I was transferred to Company H the next day and then sent to the hospital." Hearing that the regiment was being sent into combat, he appealed to his new commander, John Gossett, who agreed to sneak him aboard the *Zeilin*.[5]

On October 18, Colonel Baglien recorded that the regiment's "front line positions are well dug in, and guns are properly placed." Later that night, news spread of a major change in the command structure back on New Caledonia. Nimitz had decided that the command in the South Pacific needed more forceful leadership. So he replaced Ghormley with Vice Admiral William (Bull) Halsey, everyone's idea of a tough fighter. "We are whooping it up and turning cartwheels in the dark," a marine colonel noted. Hopes rose when Halsey summoned Vandegrift to New Caledonia a few days later. The Navy had fared poorly in the waters about Guadalcanal so far. Now maybe that would change.[6]

General Vandegrift had used the army regiment's arrival to reorganize his defense, shifting some of his own men back into reserve where they could recover from illnesses and fatigue, while putting the soldiers into what he hoped would be quieter parts of the line where they could gain some experience. As he reported after the war, he left the defense of the beaches to the "3rd [Marine] Defense Battalion, units of the Special Weapons Battalion, AmTracs, Pioneers and Engineers." The rest of the defense perimeter started at the Ilu River on the east, where the 164th's First Battalion was in place. The line then ran south, linking into Timboe's Second Battalion, turned west where Company E was in place. There was a small gap between Stannard's platoon and the lines held by two battalions of the 7th Marines, which stretched 2500 yards further west to the Lunga River, then across that river and through jungle back before running north toward the beach. Here the remaining 1st and 5th Marine regiments were dug in (see Map 2). Vandegrift was confident that this defense could hold against most Japanese attacks. Believing that the northwest corner of the line was the most vulnerable spot, he advanced some of his marines across the Matanikau River. He hoped that these outposts, "supported by artillery and 75mm [gun] halftracks," would give him some defense in depth against a sudden Japanese thrust at Henderson Field. Above all other considerations, he had to hold the airfield. Without the air support as a deterrent, the Japanese navy could sail right up to the beach and overwhelm the American position.[7]

The early days on the defensive perimeter were tough ones for many of the still-green soldiers. The Japanese navy continued to shell the American positions at night. Alvin Tollefsrud recalled that "for the first five days on the island we were bombarded by the Jap navy every night and at least three airplane raids a day. There was a reporter from the east coast, I believe his name was Dowling, who said 'we cannot write in this madness.'" Ed Mulligan remembered how many men "forgot the password or were too scared to repeat it and were shot at by friendly troops." Most guards held off firing if they "recognized a few choice swear words that should not be repeated here." Mulligan himself admitted that "with the sea behind us and the Japanese infantry in front of us," it seemed the "best choice" was to fight; "North Dakota farmers were mighty poor swimmers and no one could walk on water." Many men were reeling from lack of sleep and the effects of the massive bombardments directed at them almost every night. During one shelling, Gordon St. Claire, Sam Baglien's driver, had tried to take cover during the bombardment in a shelter filled with marines. But there was no room and he rode out the massive blasts in a piece of low ground. Two days later he learned that a shell had struck the shelter he had tried to get into, killing most of the marines.[8]

In order to collect the men wounded in the bombardments, the regiment divided the band members into stretcher teams for each company. With no band to be directed, Gerald Wright was assigned to some administrative duties. A lot of the bandsmen were still bitter that Wright "had assured us we'd never go near combat," Art Nix recalled after the war. "But the *band* was not sent into combat, *we* were sent out as individuals – it was a matter of semantics." There was nothing the men could do about it. Nix, together with his friends Art Ford and Chuck Bell, was assigned to B Company, and stayed with it throughout the campaign, seeing a great deal of combat and coming under fire more than a dozen times.[9]

Raised in agricultural settings with well-ordered fields and tended crops, the North Dakotans were fascinated by the wild beauty of the jungle growth. The coral beaches, the palm groves, and the colorful, chattering mynah birds and cockatoos were pleasant to the eye, but the rotting vegetation and the foul algae that coated the surface of the sluggish, shallow rivers assaulted the nostrils. It rained every day, sometimes in furious downpours that stung the skin, other times in steady rains that lasted for hours or days. The average total was over 160 inches a year. The humidity was ghastly. The Army tight-weave gabardine twill uniform trapped the humidity, creating conditions favorable to jungle rot, open sores, and most dangerously heat stroke. Most men went shirtless while others took to wearing the lighter weight cotton fatigues the Marines issued to its troops.

Guadalcanal's unfriendly life forms included scorpions and large centipedes which could inflict painful bites, biting ants, huge rats, crocodiles, poisonous snakes and enormous land crabs that would sometimes crawl into a man's foxhole and attack him in his sleep. One night later in the campaign, John Paulson shared a foxhole with one of his squad down at Point Cruz near the beach. "We both got nervous as we heard something shuffling over the sand toward us. I thought it would be a Jap trying to find us. Suddenly a big crab dropped in the hole and scared the daylights out of us." For every glittering tropical bird there were thousands of flies, ready to descend in swarms on the men's food. John Hagen filmed his friends fighting off the flies. He has one clip of himself, trying to eat his chow, his right hand hurriedly spooning beans into his mouth, his left hand furiously waving away the flies. Wits called this the "Guadalcanal wave."[10]

The worst fiends were the mosquitoes, millions of them, carriers of malaria and anxious to feast on exposed flesh. Despite warnings from the medics and regimental doctors to keep themselves covered, men usually discarded their shirts in daylight, and accepted the inevitable host of mosquito bites. At first many of them dodged the daily dose of Atabrine administered by the medics. Rumor had it that Atabrine caused impotence. Dr. Arthur King, the theater medical commander, filed a report predicting that "the failure to enforce the taking of the drug" would inevitably lead to a rise in malaria cases, "including black fever and cerebral malaria," which were generally fatal. He recommended that officers enforce strict "disciplinary action" to make the men take Atabrine, but admitted that patrol activity and "the constant alerts, requiring men to be in foxholes and shelters a good part of the time," would make this very difficult. An epidemic of malaria among the GIs appeared inevitable.[11]

But malaria was a relatively remote enemy. The Japanese were the immediate danger. By the third week of October, there were some 20,000 Japanese soldiers on the island, against 23,000 Americans, including the air personnel and the Seabees. The Americans had little idea where the bulk of the enemy was located, but the intelligence analysts at Vandegrift's headquarters strongly believed that the Japanese would make their main attack from the northwest, against the Marines who were dug in around the "Matanikau coastal corridor" that led directly to Henderson Field. This was exactly what the Japanese commanders wanted them to believe.

Like the Americans, the Japanese were fighting an improvised war on Guadalcanal, diverting ships, troops and resources from their campaign in New Guinea, from China and elsewhere, to regain control of their hold on the southern Solomons. Their strategy hinged on the expectation that, by seizing Henderson Field, they would force the American troops to either surrender or evacuate. They had chosen the southern perimeter, not the Matanikau sector, for their main attack. For this they had assembled some 6000 men. Most of these were part of the 2nd (Sendai) Division, a veteran outfit with a long record of victories in China and Southeast Asia. Lieutenant General Masao Maruyama, the 2nd's commanding general, had promised that his men would break through the lines and seize Henderson, in this "decisive battle between Japan and the United States." Maruyama meant to bring his men up to the American positions on the south, select a likely looking vulnerable spot, and then hammer at it with mass attacks in the dead of night until he broke through. He was confident that the Americans would not be able to withstand the onslaught of his soldiers. The attack plan took his soldiers right up to the positions held by 164th Infantry.

Maruyama did not realize that his attack plan had serious flaws. In the first place the Japanese had badly underestimated the size of the U.S. force at 10,000 men. Actually, over twice as many were there. Secondly, the Japanese planners had failed to completely neutralize Henderson Field. The Americans were always able to get some aircraft up to attack the "Tokyo Express" that brought enemy reinforcements and supplies to the island, even on the day after the battleship bombardment. As a result the Japanese lost valuable supplies and men on every run, weakening their attack forces. Because the airfield was not out of action, the commanders of the Japanese 8th and 2nd fleets, carrying more troops and supplies, remained 150 miles north of Guadalcanal. They would only advance after receiving word that the airfield had been captured. One of the naval commanders termed this problem "a pitiful example of a lack of cooperation between the [Japanese] Army and the Navy."[12]

By far the greatest drawback in the Japanese plan was the fact that they knew no more about the island's terrain than the Americans. Their elaborate plan called for Maruyama's men to be in place to attack by October 21 after marching an estimated four days, south past Mount Austen and then east, to be in place across the Americans' southern perimeter. In fact, it took six torturous days to make the brutal march through the heavy jungle. Writer John Toland, who interviewed some of the survivors of this trek, summarized their ordeals. "The winding column crossed hill after hill, rivers, streams, inching forward slowly painfully, like a great worm. Each man carried, in addition to his pack, some part of a field gun, a shell or other equipment. Since it was too dangerous to cook, all – from Maruyama to the lowliest private – lived on half rations of rice. They scaled steep cliffs with ropes, hauling up light field pieces and machine guns by sheer muscle. But by the third day the task was too much except for the hardiest, and [artillery] gun after gun had to be abandoned at the side of the trail." Dozens died of exhaustion. The march was memorialized by a young noncommissioned officer: "Trudging along under dark jungle growth, when will this march end? Hiding during the day . . . our rice is gone, eating grass and roots, along the ridges and cliffs . . . Stumble and get up, fall and get up, covered with mud from our falls, no cloth to bind our cuts, flies swarm to the scabs . . . How many times I've thought of suicide."[13]

Three times Maruyama sent messages back to his superiors that he would have to postpone his attack. The precise timetable for victory on Guadalcanal was coming apart. Meanwhile the Americans waited and wondered.

In the Company E sector of the American line, John Stannard and his platoon continued to strengthen their positions in the "coffin corner." Communications men came up and laid telephone lines back to the 2nd Battalion headquarters. There was one 37mm anti-tank gun to Stannard's right, dug in and manned by marines. Men placed extra hand grenades and extra ammunition in their foxholes. Chuck Walker had one of his machineguns in place at the coffin corner. In addition, there was a marine machine gun nearby on the right flank. Also, Walker's men and some of those in Company E had, by hook or crook, secured submachine guns, Browning Automatic Rifles and even some Winchester shotguns, wicked weapons for close-quarter fighting. Walker himself had a Reising submachine gun, a tricky weapon that required frequent cleaning but could do a lot of damage.[14]

Stannard set up a small tent behind the line where extra ammunition could be stored along with rations, canteens and canvas bags of water, and first aid supplies. Some men went out to the barbed wire laid in front of their fields of fire and tied food tins to the wire. The marines had not had barbed wire when they repelled an attack on the south perimeter in September. This new barrier, from wire brought up on a supply run, was a comfort to the GIs. They also appreciated the "kill zone" that the Marines had prepared by trimming down the long grass and foliage for about a hundred yards beyond the wire. The men called this open ground the "Bowling Alley." Company E sent daily patrols down trails and into the jungle south of the Bowling Alley, going a mile or further into the bush, looking for any "fresh sign that the enemy had been in the area." Once they found a few documents covered with Japanese script and sent these back. Otherwise they found next to nothing, some discarded ration packages and items of clothing. The patrols could be nerve-wracking. Only the narrow trails permitted movement "with any reasonable speed" and the fear of ambush was ever present. Stannard decided that "the advantage in the jungle lay with the side that did not move."

As the days crawled by, the men stayed close to their foxholes, eating, smoking, telling jokes, playing cards – and always thinking of the coming night when the dark closed in and each little noise could sound like a man creeping up to end your life. One man always remained awake in the foxhole, lest everyone died in his sleep. Behind the south perimeter, near Henderson, the 3rd Battalion was in reserve as the regiment's fire brigade.

Company L was part of this reserve, which included Sergeant Alvin Tollefsrud's squad. He was worrying about how he should place his men once they did go up to the line. "I had one fellow; I thought he was a little bit slow. I decided I'd put him in a foxhole with another guy, a big, red-haired guy, pretty muscular, I thought he looked pretty good and steady."[15]

At 3rd Battalion headquarters, John Hagen and his pals waited "in bivouac, back behind the front line." Hagen had had a premonition that the Japanese thrust would come at the 164th position, a feeling that may have been influenced by his encounter with death earlier on the Matanikau River front. Sent there to help the marines lay communications wire, Hagen had sought shelter during a Japanese mortar barrage. "I jumped into a foxhole that was kind of hidden. There were two holes, spread out kind of fan-like, at an angle, like a V. I jumped in and there was a dead marine. In the other hole there was another dead marine." Marines from the grave registration unit came forward and hauled out the bodies, but the image stayed with Hagen. "That's kind of weird crawling in with a dead marine. They'd both been shot and one was bayoneted." Now, he and his friends speculated on where the attack would begin. "We didn't know where they were."[16]

Day after day, Sam Baglien recorded in his diary that there was still no firm information on the enemy's whereabouts: "patrols still report negative, though some enemy movement is detected by our aircraft. Sector front is quiet." Two days later, Bryant Moore decided that the regiment command post should be moved. Since arrival it had been at the airfield, in "sturdy dugouts" and "well camouflaged." But the continual bombing and shelling had made the situation, as Baglien noted, "a little too hot. We discovered that 60 bombs and 50 shells had hit in our area since taking over. Our new place doesn't look too good, but it hasn't been discovered." The tension was mounting. The intelligence section's logs recorded every patrol sent out to find the enemy, only to summarize the situation with the words "negative report" again and again.[17]

On the west flank, three Japanese tanks rolled out of the trees at the mouth of the Matanikau River during the night of October 20. The marines quickly drove them off, but the action focused attention to the Matanikau front. One of Vandegrift's patrols had captured an enemy map that suggested attacks on all fronts, west, east and south, but his intelligence officers discounted this, noting that in the south nothing was seen but "bands of dispirited, half-starved, poorly armed stragglers" left over from the failed Japanese attack in September. The signs strongly suggested that the attack must come from the Matanikau sector.[18]

That seemed confirmed on the night of October 23-24, when a battalion of Japanese infantry, accompanied by eighteen Type 95 tanks, came charging at the Marines entrenched along and south of the river. Marines broke up this attack handily, using machine guns and artillery fire to cut down the infantry and anti-tank guns to knock out over half of the armor. The wrecks of the tanks and enemy bodies littered the sand spit the next morning. The Japanese commander of this attack later claimed that he had launched this futile assault as a diversion. If so, it worked, because Vandegrift, convinced that "all signs point to a strong and concerted attack from the west," took a full battalion from the 7th Marine Regiment, stationed on the south perimeter, and sent it to reinforce the Matanikau. That left the Colonel Lewis "Chesty" Puller with his single battalion, the 1/7, to hold some 7500 feet of defensive front.[19]

On October 24, General Maruyama, having finally arrived at his jump off point, deployed his men and made his final dispositions. The Sendai, reinforced by other troops, had nine battalions for an attack in columns. They intended to attack by using infiltration to cut off some American platoons, overwhelm them with mass attacks, punch through the lines, and then head for Henderson Field. As evening drew nigh, marine scouts reported to the division HQ that they had spotted a Japanese officer surveying the south perimeter with field glasses, and that smoke from cooking fires suggested that a sizeable number of the enemy was in the jungle south of Puller's men. But by then it had begun to rain and darkness was closing in, so the officers were uncertain what to do. Vandegrift was still in New Caledonia consulting with Halsey. The rain became a downpour and clouds blocked the moonlight, making it difficult to bring up reserves.[20]

At about 2130 hours, having slogged through the wet jungle, frequently struggling across bogs filled with water, the Japanese broke clear of the densest vegetation just as the rain was letting up. The men were dead tired after a week of marching. They were hungry and most of their heavy weapons had been abandoned. Now all they had was mortars and grenade launchers, some machine guns and their personal weapons. But their sprits had not flagged; they had marched the last few miles singing their national anthem and before they mustered to make the attack, they bowed in the direction of Japan, weeping and promising to give their lives for the emperor. A pamphlet issued to Japanese Army soldiers assured them that Americans were "haughty, effeminate and cowardly" and that they "intensely dislike fighting in the rain or mist or dark." A determined charge with the bayonet would break the enemy's will. The first company of infantry went forward crawling through the grass, aimed at Puller's line and at Stannard's unit in the coffin corner.[21]

The first marines to see the attackers were part of an outpost established out in the Bowling Alley. As they called back a warning and tried to creep back to the main line, the Japanese column swept past and threw themselves at the barbed wire barrier in front the 7th Marines. The wire held the Japanese up while the marines poured rifle and machine fire into their ranks. With the first group destroyed, the Americans faced a second wave, then another. The 7th had a surplus of machine guns on hand, a gift the Navy had left behind, mounted on a hundred armed landing craft. The marines seized on these as "manna from heaven!" and emplaced them across the perimeter. Now these guns would save the day, cutting down wave after wave of fanatic charges. As the most recent biographer of Puller noted, the "courageous but foolhardy Japanese simply dissolved in the face of this overwhelming firepower."

At one point the Japanese dead piled up against the wire and literally pressed a section of it to the ground, allowing some of the attackers to penetrate past the defenses. These infiltrators went to ground in the trees and brush and may have inspired the false report that Sendai divisional headquarters received, to the effect that the Japanese "had Henderson Field." The Sendai immediately forwarded this to its army headquarters, which in turn signaled to Rabaul that they had "captured the airfield." This was completely untrue. Puller, meanwhile, was moving about his lines, telling his men repeatedly that "there was no such thing as falling back." Machine guns were overheating as the water in the cooling jackets boiled away. In one spot, Sergeant John Basilone, in charge of a machine gun section, told his men to urinate on the guns and keep firing. As ammunition ran low, Puller shouted to his men, "You've got bayonets, don't you?" They fought on, taking casualties steadily. Massed fire by the American artillery helped slow the Japanese attack. The commander of the artillery reported after the battle that the combined firepower of howitzers and mortars "ended in the virtual annihilation" of entire enemy units.[22]

Whether by chance or calculation, the Japanese attack had struck close to the junction between the 7th Marines and Company E of the 164th. One column of the attackers, moving further east, came up against John Stannard's platoon and Chuck Walker's machine gun. Stannard had received a bit of advance warning near dusk when "six Native Island Scouts," who were inhabitants recruited to act as guides for the Marines, came into the lines and reported seeing about "2000 Japanese back in the hills" to the south. Stannard's platoon went on full alert.

After dark, Sergeant Lawrence Poe joined him. Poe had been with the marines in the outpost in the Bowling Alley, and had fired a couple of shots at the Japanese while falling back. Stannard, who had seen Poe compete at army rifle matches and considered him "the best rifle shot in Company E," was glad to have him. He sent Poe over to the end of the position to provide cover for the 37mm gun. "Rapid fire from [Poe's] M1 helped to turn the enemy back at one of the main points of attack," Stannard reported later. The sergeant directed other riflemen to target enemy "machine guns [out] in the field [placed] to fire on the Company positions."

Walker's men were in the thick of it as well. Walker was going over details with his platoon sergeant when he heard his machine gun open up. Within seconds, M1s and the 37mm gun were also in action. Walker scurried over to his closest gun crew, handing out more belts of ammunition. He stopped to talk to his gunner, Kevin McCarthy, and asked him what he had seen. "Japanese came along the edge of the bush in a column of two, straight at us," McCarthy answered. "When they reached the barbed wire they stopped in confusion. When I yelled Password! You should have seen the cigarettes fly. Heck, most of them were smoking." McCarthy kept firing as fast as his loaders could feed the gun. Walker would always remember the friendly advice that McCarthy, a "good Dakota farm boy," gave him: "the bullets are thick as flies on horseshit, watch yourself!" Walker said later that without the fast-firing M1s that the Company E men had, "we'd have been overrun" while letting the machine gun cool down.[23]

Behind the lines, the 164th's heavy 81mm mortars also joined in, dropping barrages as they were called for by forward observers. Don Hoffman began hauling shells over to his mortar crews in Company H. "We had two kinds of shells, a smaller round that we could fire a long ways, and a bigger round that had less range but could really do a lot of damage. The Japs were trying to push down the wire, so we used the big shells just beyond the wire and really tore them up. We had a truckload of shells when it started and used all that and another few truckloads by the time it stopped. I don't think more than two of our tubes were any good to us after the battle, they had gotten so hot."

Al Wiest, the commander of Company M, had been concerned that the mortars would be hampered by the dense jungle and the contact fuses on most of the mortar shells. "The heaviest [81mm] mortar shells were the only ones with a delayed fuse. Our lighter 60mm shells and the smaller 81mm shells didn't have those delayed fuses. You had to blow half of the jungle canopy away before the darn things would land down on top of troops. But that didn't take long to do the way we were firing," especially as the range closed. "We were pulling those 81s in close, where they were bursting 50 yards in front of our own men. They committed mayhem." Sam Baglien reported after the battle that "Company M's mortar platoon fired 1200 rounds of 81 mm stuff at the Japs" on the first night of the battle. The 60mm mortars attached to each rifle company were also in action. When all the other men in his 60mm crew were killed or wounded by enemy counter fire, Sergeant Joe Otmar worked the tube by himself, firing over 500 rounds. The mortar barrages, combined with the 75 and 105 artillery bursts, turned the ground in front of the barbed wire into a slaughterhouse.[24]

From time to time, Japanese snipers who had slipped past the front line, would try to pick off the mortar crews. "One of our guys got a Thompson and went out looking for them," Don Hoffman recalled. "I can still see him, wearing nothing but a pair skivvies and his boots. You'd hear a shot from a Japanese rifle, they sounded different from ours, and then you'd hear his Thompson open up."

Over in the 7th Marines part of the perimeter, Puller had no reserves left for his single battalion. The Japanese had pushed a bulge into his line. He had already called back for Bob Hall's 3rd Battalion to come up and enter the fray. Colonel Moore ordered two platoons from Company G and one from E, held in close reserve, to move up quickly while the remainder of the 3rd Battalion was alerted. Harry Wiens was sleeping in his trench with the rest of Company I when "firing broke out of the southeasterly side of the perimeter. That was opposite from where it had erupted the previous night." Soon after, the call came from regimental HQ to form up. "By that time we were already standing about with weapons in hand." At the battalion HQ, Hall was calm and, as Baglien and Al Wiest remembered, "all business." Hall "issued his instructions to his battalion [staff] and immediately went forward on a personal reconnaissance." The battalion began marching toward the front line, arranged in single file on the trail. In Company L, Alvin Tollefsrud gathered his squad and put them into the file. "You had to hold on to the guy in front of you, it was so dark," he recalled. The long hike gave him a lot of time to think: what would he see when he got into the fight, how would he respond, would he die up there, tonight? On the way, Tollefsrud and his men passed the marine artillery gunners, who shouted encouragement as they marched by: "good luck guys, we'll give 'em hell, all the [fire] support you want."

As Company I joined the trek south, company commander Ralph Knott sent Wiens with several of the platoon leaders to get information from General Roy Geiger, the overall commander in charge while Vandegrift was in New Caledonia. Geiger explained the situation to the Army lieutenants. Wiens heard Geiger say that the battalion had to stop a "breakthrough" to the airfield. He thought Geiger looked very worried. Then the GIs piled into a jeep and, with Wiens riding "as shotgun" against possible snipers, they went up to the perimeter. As the jeep passed the companies marching south in column, Wiens was surprised at the "remarkably good time" the men were making "in that mud." Arriving at a 7th Marine command post (CP), Wiens and the platoon lieutenants started arranging to guide the GIs right into the line alongside the marine platoons. Suddenly an "excited Marine guide returned to the CP with a firm pronouncement that he was going to get himself an M1, even if he had to steal it." While guiding some of the first GIs into the line, he had suddenly run into "five Japs" coming out of the brush. He shot one and the army sergeant "dispatched the other four before he could retract his bolt and chamber another round." Wiens was amused when the young man said "he was absolutely going to get one of those rifles!"

Soon after this, Colonel Hall arrived and met with Chesty Puller. Puller told Hall he wanted to feed the GIs into the positions by platoons, mixing them among the marines in order to fill gaps in the line. Hall simply said, "That's fine with me." It took about an hour, from

0230 to 0330, to file the 3rd battalion men into empty bunkers and foxholes. Much of the guidance for this came from Father Matthew Keough, the 7th Marines' Catholic chaplain, who led men into positions where reinforcements were badly needed. Virtually all of the battalion went into the line. "The 3rd Battalion [had] but two squads in reserve, as all troops were in the line," Baglien later reported. Japanese companies continued to attack. Armed with the M1, the 164th men quickly made a difference; "the sound and tempo of firing picked up tremendously," a later report stated.

Back behind the fighting, a last ditch defense of the airfield was being organized in case the Japanese did break through. The Seabees were alerted to take defensive positions with their pistols and light M1 carbines. Even the medical personnel and some of the wounded found weapons. No one expected the Japanese infiltrators to take prisoners. The remnants of the regimental HQ, which Baglien summarized as "the Service Company, riflemen from the AT [anti-tank] Company, and what runners and odd men were around the regimental CP," assembled. About 175 men in all, this group included Doug Burtell, who remembered that he and the others were ordered to hold the airfield and vital installations. In retrospect, he noted, "we didn't have to do anything." At the time, they waited for the Japanese to come charging out of the night.

A handful of men left back at the regimental HQ manned the telephones and received messages from runners. They were shaken when Stannard's platoon reported turning back a bayonet charge, but relieved to find that the other battalion sectors were relatively quiet.

Of course, in those sectors men's nerves were stretched tight. At Company F, just to the left of E Company, John Paulson and his men peered out beyond their line, tense in expectation of an attack. "We were in our holes, hearing the battle, waiting for a bunch more [Japanese] to come at us." Walter Rivinius, also in F Company, remembered that the machine gunner to his right "let go [a burst] because he thought he saw something. Our sergeant yelled out, 'Hold your fire, hold your fire.' Who knows what it was, we weren't going to go out and look around." Down at the beach, Dennis Ferk clutched his M1 and waited with the rest of Company A. "We had barbed wire in front of us, with cans and tins tied to it to make noise. We heard something rattling the cans and we all fired. Next morning all we found was a dead dog."

Between 0400 and 0500, the Japanese thrusts began to slacken. Both sides were becoming exhausted. "In some cases hand to hand combat for the possession of foxholes and emplacements [had] occurred," Sam Baglien noted. There were numerous cases of individual heroism. Louis Lochner's Company G platoon was among the first men to reinforce the marines. He engaged several charging Japanese that were closing in on a marine machine gun emplacement, killing at least six and driving off the rest, even as he was badly wounded. Lochner subsequently died. Bill Clark was a corporal in Company M, in charge of a heavy machine gun squad. One of his officers remembered Clark as a "sandy-haired, quiet, good sense of humor . . . dependable type" guy, who had doggedly taught himself on New Caledonia to take his Browning .30 cal. gun apart and reassemble it – blindfolded. Now, fighting in the 7th Marine lines, his gun had jammed, as had another .30 caliber gun a few yards away. Clark proceeded to take "undamaged parts from each, and reassembled them into one operable machine gun," in pitch blackness, "while casualties were occurring around him." The rebuilt gun, back in action, played the pivotal role in repelling the next Japanese wave. Both Clark and Lochner were subsequently awarded the Distinguished Service Cross.

John Stannard's and Chuck Walker's men had littered the coffin corner with Japanese dead. Walker counted "hundreds" of dead near the wire. He later described how at one point his gunner, McCarthy, and other men went to the aid of some of the Marines who had withdrawn from the listening post in the Bowling Alley. "Bob Campbell of our Company H had brought up a [Bren] carrier to rescue the Marines still mixed with the Japanese in the high grass. McCarthy and two other men jumped into the carrier, which ran across to the Japanese line, defying the odds. Armed with only one machine gun on the Bren, McCarthy held off the enemy, enabling the Marines to pile in." Miraculously, the carrier and all its occupants made it back unharmed. McCarthy would also be decorated.[25]

There were many heroic acts that night, a lot of them not witnessed or unrecorded. One that was recognized was Lester Kerbaugh's. During the fighting, Kerbaugh and his rifle squad had extended their positions to the right in order to close some of the gap between Company E and the 7th Marines. The Japanese had subsequently surrounded the squad and launched several attacks to overrun the position. In the early morning hours of the 25th, in order to help spot the enemy in the thick foliage, Kerbaugh "deliberately exposed himself to the enemy and taunted them, in order to draw their fire." As five members of the squad testified after the battle, Kerbaugh "stood erect" and shouted insults at the Japanese, daring them to hit him. By spotting the rifle and machine gun flashes, the rest of the squad killed a number of Japanese and the squad was then able to fight their way back toward the rest of the Third platoon without loss. "I feel that the actions of Private Kerbaugh saved the squad," concluded one of the men. Kerbaugh was awarded a Silver Star for his daring exposure to fire.[26]

Men fought desperate little battles in the dark, their senses barely registering the snap of the bullets, the flashes of the explosions, and the screams of the wounded and dying as their whole being focused on staying alive. They fought and killed and bled, beating back the attacks until after dawn, when the Japanese began to withdraw back into the jungle. "Come dawn," Harry Wiens wrote, "I sat down in a corner of the hole and slept a little, as the firing to our left had slackened though not completely abated." Al Wiest was at Puller's command post when he heard Chesty remark to someone on the field phone "I'll tell you one thing, these farms boys can fight." Sons of pioneers of the Midwestern prairies, children of the worst economic crash in the American experience, the men of the 164th had shown they could stand and face the enemy. In a scant few hours, they would have to prove it again.[27]

(Endnotes)

[1] Engstrom diary, entry for October 22, 1942.

[2] The Moore incident is related in Frank, *Guadalcanal*, p. 313.

[3] Timboe information from his daughter, Mary Fran Riggs; John Stannard's memoir, *The Battle of Coffin Corner and Other Comments Concerning the Guadalcanal Campaign* (Gallatin, TN: privately published, 1982) is a vital source for Company E's experiences in the coming battle.

[4] Stannard comments frequently on Kerbaugh in his memoir. General Charles Ross (ret.), in an interview with Blake Kerbaugh (Lester's son) made the comment about Kerbaugh. My appreciation to Blake Kerbaugh for a copy of this interview and for providing the story of how his father joined Company E.

[5] Charles Walker, *Combat Officer*, pp. 6-8, along with Shoptaugh interview of Walker, March 2009.

[6] Baglien diary, October 18, 1942; Miller, *Guadalcanal, The First Offensive*, p. 170; Merillat, *Guadalcanal Remembered*, p. 183.

[7] Stannard, pp. 28-29; Walker, pp. 12-14; Vandegrift, *Once a Marine*, pp. 182-83. Frank, p. 343, notes that the "novice 164th" men were placed in the "easternmost sector, the least likely subject of enemy activity." An "AmTrac" was a lightly armored amphibious tractor armed with several machine guns, a powerful weapon when dug in.

[8] Alvin Tollefsrud, interview; Edward Mulligan, "October 13, 1942," *164th Infantry News*, November 2002.

[9] Art Nix, undated [1990s] tape memoir, recorded and given to James Fenelon, who provided a copy to Shoptaugh.

[10] John Paulson story, told to Shoptaugh in July 2009; John Hagen tapes and movie film, NMHC.

[11] King, "Intelligence Report on Cactus [Guadalcanal]," October 15, 1942, Arthur King Papers. In another section of this medical report King raised the touchy issue of "war neurosis." He wrote that he found many men "in a pitiful state of nervous exhaustion after the long period of combat without relief," pointing out that Japanese aircraft had bombed the men 61 times in 68 days. He felt most men would recover with some rest and hot food. The official Army study *Medical Service in the War Against Japan*, by Mary Ellen Condon-Rall and Albery Cowdrey, (Washington DC: U.S. Army, Center for Military History, 1998), p. 126, notes that General Patch was inclined to reject war neurosis as genuine and at one point "insisted that all cases should be court-martialed" for cowardice. Doctors took to diagnosing fatigue cases as "blast concussions" to prevent a rash of prison sentences.

[12] See Raizo Tanaka and Roger Pineau, "Japan's Losing Struggle for Guadalcanal," *US Naval Institute Proceedings*, (August 1956), Frank, pp. 337-42. Several 164th veterans commented that they never understood why the Japanese did not place their heavy artillery on Mount Austen, which would have given them an excellent view for shelling the airfield. But the Japanese lacked animals or vehicles to haul ammunition for heavy guns too far into the interior of the island and had only what Frank, p. 339, calls a "small hoard of artillery ammunition."

[13] John Toland, *The Rising Sun* (New York: Random House, 1 vol. paper edition, 1970), pp. 454-55; Warrant Officer Yosida Kashichi, "When Will This March End?" in Saburo Ienaga, *The Pacific War* (New York: Pantheon Books, 1978), p. 144. See also the comments of William H. Whyte, *A Time of War* (New York: Fordham Press, 2000), pp. 76-77, and George W. Smith, *Do or Die Men* (New York: Pocket Books, 2003), p. 166, both quoting captured documents.

[14] Stannard, pp. 60-68, which covers this first day of battle. See also Walker, pp. 13-17. While in his book Walker notes he had four water cooled and four air cooled machine guns in his platoon, he notes in his interview with James Fenelon and his interview with Shoptaugh that he had just the one .30 caliber gun attached to Stannard's riflemen. The rest of the guns were in other parts of the Company E sector. The Browning Automatic Rifle (BAR) was a clip-fed light machine gun.

[15] Tollefsrud interview with Shoptaugh.

[16] John Hagen tapes, NMHC.

[17] Baglien diary entry for October 21, 1942; S2 Journal entries for October 20-23, 1942.

[18] Miller, p. 156.

[19] Miller, pp. 156-57. The '1/7' is the proper designation for 1st Battalion, 7th Marine Regiment.

[20] The events of October 24-25 on Guadalcanal are related at length in Frank, pp. 339-57; Cronin, pp. 53-55; Miller, pp. 160-62; Cooper and Smith, pp. 280-84; and Hough, pp. 330-35. Several quotes of the actions on this night are taken from John T. Hoffman, *Chesty: The Story of Lieutenant General Lewis B. Puller, USMC* (New York: Random House, 2001).

[21] In an undated paper, "The 164th on Guadalcanal," Colonel Arthur Timboe, commander of the 2nd Battalion during the Henderson battle, notes that the term "coffin corner" was coined after the battle because of the large number of Japanese killed in front of Company E's position (copy of paper courtesy of Timboe's daughter, Mary Fran Riggs).

[22] Colonel Pedro del Valle, "Marine Field Artillery on Guadalcanal," *Field Artillery Journal* (October 1943). In this article, del Valle stated that he believed

an entire Japanese regiment was destroyed in this first night of the battle. Frank, pp. 354-57, suggests that because of the difficulty the Japanese commanders had in getting some of their attack columns in line, the losses were not this high.

[23] In addition to Stannard's account, the personal memories for this chapter are from Walker, *Combat Officer*, pp. 18-20; the interviews of Douglas Burtell, Dennis Ferk, Don Hoffman, John Paulson, Walter Rivinius, and Albert Wiest; the SHSND interview of Charles Walker; Samuel Baglien, "The Second Battle of Henderson Field," in the *Infantry Journal,* May 1944; and Harry Wiens, "My Little Corner of the War," pp. 88-92.

[24] Otmar's actions are reported in Cronin, p. 55.

[25] Decorations for the Guadalcanal campaign are summarized in the 164th HQ "Report of Battle for Henderson Field," June 28, 1943, 164th Infantry Association Records, UND. See also Cooper and Smith, *Citizens as Soldiers,* pp. 54-55; Walker, *Combat Officer,* p. 21; Richard Stevens, "Bill Clark's Game, *164th Infantry News,* July 2008; and Cronin, p. 55. A Bren carrier was a small British-made, lightly armored and tracked vehicle. Some were used at Guadalcanal.

[26] "Report of the Investigation of Distinguished Service of Private Lester Kerbaugh" [November 1942], with attached recommendation from Colonel Bryant Moore, February 9, 1943. A copy of this document was obtained from Kerbaugh's Army personnel file and shared by Blake Kerbaugh.

[27] Wiens, p. 92; Puller comment taken from September 2007 conversation with Albert Wiest, who was at Puller's headquarters with Hall.

Chapter 6: "The Din Was Terrific"
Guadalcanal: October-November 1942

General Vandegrift flew back to Guadalcanal, cutting short his talks with Halsey, pleased at the Admiral's promises to give him greater naval support but a bit angry with himself for having been gone when the Japanese made their assault. "The enemy's southern attack had fooled me," he admitted. "Before I left Guadalcanal I was convinced from intelligence reports that the main enemy attack would come across the Matanikau." But fortunately Puller's men held, and "the Army squads, platoons and companies" of the 164th had "more than justified their existence in the ensuing hours."[1]

Had the Japanese shot their bolt or would they come on again? General Geiger guessed that Maruyama would wait until dark and attack the south perimeter again. He rejected suggestions from his staff to launch a counterattack and told Puller and Hall to strengthen their positions for another attack. The Japanese artillery resumed blasting at the airfield at first light, and enemy bombers began pounding Henderson again. At one point, enemy destroyers sailed boldly up and lobbed five inch shells into the perimeter. Shore guns returned fire and scored at least one hit, forcing the Japanese ships to withdraw. After the wet ground at Henderson Field dried enough, American fighters went after the Japanese, shooting down over twenty planes. The Japanese artillery continued to fire however, and Zero fighters made numerous strafing runs over the American lines. Under the volume of all this fire, the First Marine division HQ advised units that the "safest course is to lie low in daytime."[2]

The Americans called this "Dugout Sunday," but the image is somewhat misleading. Certainly many did stay in their shelters. A lot of men were very busy and had no choice but to brave the enemy fire. Out in front of the wire, Bill Welander and others risked sniper fire to search the Japanese dead for diaries, maps, letters and other items of intelligence value. Welander found it gruesome to be out there, in the midst of charred bodies and human waste that overwhelmed even the fetid odor of the jungle. The jungle grass was "thick with dried blood."[3]

Over in Company E's sector, men were still trading shots with Japanese soldiers who were sniping at the Americans. Owen Heller, in John Stannard's platoon, was trying to spot snipers in the early morning light when a Japanese soldier dashed past him and tried to reach the jungle behind the American front line. Heller and several other men shot him down. A few minutes later, Heller saw a pile leaves moving near enemy bodies at the wire; he crept over and prodded the leaves with his rifle. Another Japanese soldier sprang up and ran for the brush, but Heller cut him down as well. Still later, a Japanese Nambu machine gun opened up on the American foxholes. Heller spotted the fire and saw that the gun was dug in around the roots of "a giant jungle tree." He emptied his rifle at the roots, but the gun remained in action most of the morning.[4]

The soldiers had expended most of their ammunition in the night fighting, so other men had to go back for bandoliers and machine gun belts. Lester Kerbaugh and one of the marines in the "coffin corner" started back toward a command post to get more ammunition but saw "a number of Japanese soldiers moving about" in the brush. They scurried back to the line and told John Stannard that the men would have to watch "both the front and the rear." Not long after this, Kerbaugh shot a Japanese soldier who was trying to crawl up behind him.

Richard Stevens was also busy that morning. Ever since transferring into the 164th just prior to its shipment to the Pacific, Stevens had been his platoon's "agent and instrument corporal," a catch-all job in which he used surveying instruments to determine indirect fire angles and elevations for the machine guns. Stevens was also an alternate gunner, ammunition carrier, and general dogs-body for whatever needed doing. During the previous night of fighting, he had marched up to the perimeter with the rest of 3rd Battalion, hunching his shoulders away from the "lines of incoming machine gun tracers" passing overhead, and thinking grimly about mortality. He and his friend Richard 'Slick" Zerrill were assigned to help a navy corpsman, using shelter halves and ponchos to carry wounded men back to the aid station. At some point during the long night, he heard an old marine veteran say to his mate: "This is just like Nicaragua!" After hauling out wounded men, he and Zerrill joined a makeshift group of soldiers and marines in hunting snipers. Then at first light, Stevens returned to company headquarters and used World War I-era equipment to reload machine gun belts. Extra guns were being added to the defense and Stevens' commanding officer told him that come dark he would man one of the guns.[5]

Al Wiest, Stevens' company commander, was moving along the line as well. "I was a busy man that day, checking out our [Company M] machine guns, making sure they all had enough ammunition. I was carrying out my headquarters duties from my back pocket." Dodging the occasional artillery shell or bomb, Wiest visited each gun site to assess the condition of the weapon and make certain "there weren't any dead spots" along the fire lanes. He was worried that the water cooled .30s might not hold up to a second night of intensive fire, so he ordered extra water brought forward to cool the gun barrels. By the time he returned to his HQ shelter, it was already growing dark. Then all he could do was worry. "I envied the rifle company commanders; they could do more once the shooting began." Somewhere nearer to the 7th Marines, John Tuff and John Kasberger were manning their own machine guns. Kasberger, a cheerful fellow who, like Stevens, had transferred in from the Kansas National Guard, was musing that "October 25 was my birthday and I couldn't help thinking that they sure had a lot of fireworks for me that [first] night!" Like every other gunner he worried his gun would overheat once the water jacket had been depleted. "But during a lull, some of the steam re-condensed back to water so we got by. We had plenty of ammo." He was reassured to know John Tuff was close on hand. "John had more common sense than most people and thought things out before he made a move. We felt safer around him."[6]

The rifle company officers and noncoms were also out and about during the shelling and bombing, reorganizing the lines. The line of the 7th Marines, 1st Battalion was now shortened, giving it a smaller section to defend on the eastern part of the perimeter. I Company of the 164th extended west from the Marines left flank, followed by K Company and then L Company which linked its left flank to the Coffin Corner. Bob Hall's battalion was now in place, his men looking out over the low-lying ground that was covered with hundreds of Japanese dead. Hall placed his 60mm mortars a few hundred yards behind the rifle pits and had them register their tubes to drop shells just beyond the wire. The 81mm mortars were further back, their registrations placed on the edge of the dense jungle. Using machetes and captured Japanese bayonets, the soldiers braved the snipers to chop down more of the kunai grass. The "killing field" now reached out from 60 to 100 yards in front of the fire pits. Expecting the Japanese to try and break through the junction of the 2nd and 3rd battalions, Hall placed 37mm guns to cover that spot in the line.[7]

Alvin Tollefsrud was getting his L Company platoon into position and his day started out with a close call. "That morning me and another fellow got up [out of our holes] to go out and look at the dead and as we walked out a marine called to us 'watch it!' He looked like he wasn't more than fifteen years old, but he had that 'thousand yard stare' already. All of a sudden he shot a Jap officer who had got up from among the bodies out there and was coming toward us. Stanley Anda, one of our old sergeants, then took a submachine gun and went out there shooting any Jap he saw move." Tollefsrud's squad was very close to the junction point with E Company and he was happy to see the 37mm guns so close. As he placed his men, he made certain to pair the man he thought "a bit slow" with the redheaded, muscular fellow he thought looked "pretty good." Tollefsrud placed himself in the center and slightly behind his platoon line, where he could see everyone and direct some of the action. "So we reorganized our line. And then they came that night, they really came."8

Japanese artillery fire continued on and off all day. The enemy bombers struck one more time before it got dark, and the artillery fire began to concentrate on the south perimeter. At just about 10:00 p.m., Japanese in the jungle also began calling out insults, apparently in an effort to draw fire so they could mark American positions. This lasted for about an hour and then a series of red and green flares went up from the enemy positions. Then came the charge. This time Maruyama arranged his two regiments closer together and sent them headlong at the American battalions, the enemy soldiers screaming curses as they ran at the wire. Once again, part of the Japanese force did not hit the wire as planned, its commander later explaining that he held back to prevent "an American attempt to move around his left flank." The 16th Regiment of the Sendai Division, making up Maruyama's right wing, threw most of its power at I, K, and L Companies in Hall's battalion, and at Coffin Corner.9

The attack was greeted by some as a relief. All day long they waited, checked their weapons, got some chow and thought about the coming attack. Now it began. "They came at us from out of the tall grass, and some got through the wire," Alvin Tollefsrud recalled. Trees in this part of the line made it difficult to use hand grenades. Fortunately the "E Company had a 37mm anti-tank gun firing canister and it shot into the Japanese flank. I think if it hadn't been for that 37mm gun, a lot more would have gotten through [the barbed wire]. Bill Krogh was behind one of our machine guns and he saw them coming. He pulled the trigger and kept firing until that belt had gone all the way through. Then he slammed another belt in and kept firing. Pretty soon the barrel began to get red and all of a sudden it froze up. We were firing with our M1s. Some of them [the Japanese] had gotten over the wire and were coming right at us. I remember one, he was coming hard, and I don't know who shot him, but he was hit in the eye and about a quarter of his head was blown off. Normally when you hit someone like that they fall backwards, but he was coming so hard he fell forward just a few feet away from us. There was a lull after that charge, and I remember one of guys made a joke. He called out 'pick up your brass.' That got a laugh. Then the Japs came again." At one point a volley of 81mm mortar shells hit Tollefsrud's line, killing one of his men and nearly killing him. "Now I got this second hand later, but I heard those were our shells, fired because someone had been told the Japs had broken through our line. But we held. It was a pretty long night."[10]

Over in E Company, the men were pouring fire into the Japanese trying to break through at the battalion junction. "Large volumes of American artillery and mortar fire were delivered against this area," John Stannard wrote gratefully. His riflemen now were supported by BARs and a couple more heavy machine guns. They resorted to grenades when the enemy got close. Stannard told his men to be careful with the grenades, which "had an eight second fuse at that time." Hold the grenade for "a three or four second count" before throwing it, he said, or the enemy could throw it. "It was important that the thrower not stutter or lose track of his count," he wryly admitted. Charles Walker, staying with his machine guns in this section of the line, noted that with all the artillery and mortar fire, and the anti-tank guns using canister shot, "the din was terrific."11

About three hundred dead Japanese were later counted at this part of the battlefield. Some of them had been cut down by a platoon of G Company, sent south by the regimental headquarters as night fell, to fill in the small gap between L and E Companies. Neal Emery, the platoon sergeant, sent his men into vacant foxholes left by the marines. "We had not been briefed about how dark it could get, but it was absolutely black that night." The platoon's BAR man Roland Dahl took position in the center of the line and was soon busy driving back the enemy attackers. "The Japanese kept coming and coming and they were piled in front of him," Emery remembered. Firing his own rifle at any movement to his front, Emery kept smelling "ether or something." In the morning he found "five dead men within ten feet of me, one with his head wrapped in a bandage and some kind of medicine he had used, that was the smell." Another Company G man later noted that the Japanese seemed to run directly into the barrels of the Americans' guns: "it was just like shooting ducks with their wings clipped."[12]

As the battle continued, many of the telephone lines were damaged by Japanese grenades, so Don Hoffman went up to the front line to see for himself if his mortars were on target. There he got a look at the carnage first hand. "There were Jap bodies piled up against the wire. Others were back in the grass, firing and yelling 'blood for the emperor!' And one of our guys yelled back 'yeah, and we're piling 'em up for Eleanor!'" Japanese mortars had severed a line from L Company headquarters to 3rd Battalion. A runner was sent back to get a communications man to fix the line. John Hagen was in the HQ dugout when Bill Boyd walked over to him and said, "We have to get a heavy line down there." Hagen replied that all the others were out handling other line problems and he could not haul a reel of heavy line alone. "No," Boyd said, "go find the break in it and fix it." So Hagen set off with his tools and a pistol, crawling along the line to find the break. "It was so damn dark. That crap [shrapnel] was flying through the air like you wouldn't believe. The Japs would throw up a flare and you had to freeze right to the ground." The smallest movement could draw fire from friend and foe alike.

About half way to the L Company dugout, Hagen approached a group of trees and saw a shadow crouched against one of trees. "I saw that helmet and I knew it was one of our people unless a Jap had put on one of our helmets." Gripping his pistol and crawling closer, he recognized Rilie Morgan who he had met in college. He called Morgan's name in a low voice. "Who is it?" "It's Hagen, I'm looking for a break in the wire." "Oh, John, get up here. I want to show you something." As Hagen got into the trees, he saw a trail that led back into the jungle. In the light of shell bursts and flares, he saw what looked like a pile of rubbish on the trail. "Those are all Jap soldiers," Morgan said, "I sit against this tree and pretty soon another will come out on the path." Hagen watched him "dump a couple" more men. Then he advised Rilie to move on before the enemy figured out where he was. "Yah," Morgan said, "I think they're going to get wise to me pretty soon."

So Morgan led Hagen to the company HQ and Hagen found the break a few yards behind the dugout and spliced the line. Then he went into the dugout to verify that the phone worked. Suddenly a Japanese soldier who had gotten close enough to use a grenade launcher fired a grenade that landed outside the dugout entrance. "It threw dirt all over our backs and down into the dugout." Shaking that off, Hagen went back to 3rd Battalion. Later on, he went out to repair a phone unit at one of the 37mm gun positions. While he worked on the phone, he watched the guns pour canister into the enemy ranks. "They kept coming until they were all mowed down. The [enemy] officers were waving their swords over their heads and screaming. They hit us the same way they had the night before. Think they'd learn, wouldn't you?" Over in the line of I Company, Harry Wiens also marveled at the enemy's stubbornness. The light of the 37mm gunfire gave him just enough light to see "the charging legs of the Japanese as they struck at our wire . . . our artillery [support] was magnificent. They laid their rounds just forward of our wire and tore the Japs to bits."[13]

Somewhere in the Company L positions Dick Stevens was manning his machine gun. Coming up to his gun, Stevens had passed the dead body of Ira Woodall, one of the men he had carried wounded with the night before. That shook him a little, but once the attack began, he was too busy to think about it. "We were scared to death, but we had been trained well and were too busy to think much past the need to keep the gun in action. I wasn't worried about ammunition because we had a lot of that, but I was concerned about overheating the gun. If you fired too fast, you had to change barrels, but we didn't have a spare barrel. We also didn't have any spare water for the cooling sleeve. So we had to fire carefully." Stevens saw very few Japanese up close. "There was double-apron barbed wire out in front of us and that saved the position. Beyond that there were a lot of Japanese, but all you saw were flashes that looked like a bunch of fireflies. We just fired at the flashes. Every weapon was firing as fast as we could. A few infiltrators were the only Japanese you actually saw." All night long, Stevens found the heavy BOOM of the 37mm canister rounds to be "reassuring." One of the men in the rifle platoon next to Stevens was a burly fellow from Minnesota named Russell Campbell, who everyone called "the boar." Campbell "got out of his foxhole to take a piss and was shot in the leg by another soldier. He called out 'hey, it's me, the boar,' and immediately another guy shot him in his other leg! He got off with superficial wounds."[14]

West of the soldiers the Japanese struck hard against the 2nd battalion, 7th Marines, at a point where the line was thinly held by just one platoon. Taking heavy casualties, the enemy managed a brief breakthrough. However, a counterattack led by the battalion's executive officer drove the enemy back, killing some 300 more men of the dwindling Japanese force. The attackers began going to ground and the assault petered out with the coming of dawn. Stannard could see the "tips of these bayonets bobbing along above the [high] grass" and called down artillery fire on the retreating Japanese. When a short round hit a tree behind their own lines, they quickly called back to the guns to check their elevations; shells "don't know the good guys from the bad guys." Japanese snipers who had infiltrated past the wire remained active, and men had to move around with care. One man was very annoyed when a sniper barely missed him while he ran back to his squad's latrine. He felt that "should be included [forbidden] in the international Rules of Warfare!" Company F sent some of its squads over to help hunt down snipers, and later in the day Airacobras lifted off Henderson and flew strafing runs to suppress snipers.[15]

With the Americans holding fast, the attacking Japanese were again decimated. Looking at the piles of bodies stacked against the barbed wire or laid out in long lines where the machines guns and canister shot had mowed them down, men thought the dead looked more like ragged children than would be conquerors. Maruyama had promised a victory in this "Decisive Battle between Japan and the United States," but after two nights of trying he had failed. The battle was essentially over at this point, although on the next night, American machine gunners fired at groups of Japanese spotted once again beyond the perimeter. Chuck Walker noted that these last "attacks were sporadic and failed to make headway." Japanese records of the battle suggest that most of their men out that final night were either trying to recover wounded or infiltrators who were trying to get back to their units or men in search of water. Riflemen killed a Japanese soldier near the company command post that night, found he had two empty canteens and thought he must have been trying to get to the post's Lister bag.

After three nights of fighting, Maruyama's force was spent. He reported later that he had "not a single reserve left; no food was to be had nor any to be expected." He decided it was "impossible to tear the enemy positions," and ordered a retreat. He sent one body of survivors east toward the beach where Japanese destroyers might meet them with food and ammunition. This unit, commanded by Colonel Toshinaro Shoji, would be engaged again by the 164th a couple weeks later. The rest of the tattered remnants of the Sendai Division started back toward Mount Austen, but a great many would die of wounds, disease and hunger before reaching it. Left behind on the battlefield were close to three thousand dead. Henderson remained in American hands.[16]

    It was not obvious that the battle was over until October 27. It was then that increasing numbers of marines and soldiers began to venture out into the field to hunt for souvenirs and in some cases stray snipers. "Hey Wiens, let's go get some Nips," a friend from K Company called out to Harry Wiens. Wiens declined the invitation, deciding he would rather rest. A little later, he heard several shots and heard his friend yell out "we got another four!" Many of the men were like Wiens, just wanting some sleep. A few were completely shattered by their experiences of combat. The "large redheaded fellow" that Tollefsrud had expected would be steady was jittery from the moment the fighting started. "He'd shout 'here they come,' and fired wildly. Soon he'd be out of his hole and creeping back." Tollefsrud grabbed him and got him back into the foxhole. Then he had to do it again. Meanwhile the "slow" fellow was methodically taking aim and firing his rifle. The big man's nerves broke. "The medics took him back and I never saw him again." "We were all scared," Tollefsrud later said, but most were more afraid of "letting down our friends than anything else."[17]

Patrols sent out to make certain that the Japanese were in fact retreating found dead bodies, debris left behind in the grass and little else. Any living Japanese found was shot immediately. "We had no trust and little sympathy for the Japanese," John Stannard commented. Tollefsrud remembered a man in his company using a submachine gun to shoot the wounded. "He never even paused, just fired and moved on." Sniper activity added to the determination of the men to shoot on sight. The snipers, in Sam Bagien's words, "caused little damage [but] were well camouflaged . . . making it difficult to locate them." When John Hagen and others went to get a meal at a field kitchen, they came under sniper fire. "I was standing in line and all of a sudden there was a burst from a Nambu [light machine gun]. It kicked up dirt right beside me." A man spotted the fire coming from a tree about thirty yards away. "I'll tell you, everybody was shooting at the top of that tree and pretty soon he just tipped down and the gun fell to the ground."[18]

Virtually none of the enemy stragglers left on the field attempted to surrender. John Paulson was part of a group that combed the jungle growth. "There was one [Japanese] who was lying under a bunch of thick brush with his gun. I motioned for him to come out, saying we weren't going to hurt him. He just looked at me and started to pick up his gun. One of the other guys shot him right away. He didn't even have any ammunition at all. He had a big wound in his side and if he had surrendered he could have lived. It didn't make any sense." Not every straggler was helpless. Howard Lauter related later that Sergeant Ray Holzworth, a friend for years, came over to see him after the battle. "He left to go back to his area, when we heard a shot. Later we found out that a sniper had killed him."[19]

On the night of the 29th, a small enemy group came out of the jungle and set off a makeshift flare with a bundle of explosives mixed with kindling (or perhaps flares). In the light, they attacked Rilie Morgan's platoon. Twice Morgan crawled out under close fire and used blankets to smother the illuminating blaze. For this he was later awarded a Silver Star. By then the Japanese corpses were posing a hazard, so burial teams were organized to dig long trenches and throw in the bodies. Hundreds "were buried in front of the lines of K and L Companies alone." It was a horrible job, noted Bernard Scheer, a mortar team leader with K Company. "We dragged these dead Japs over to big ditches that the CBs dug and dumped them in there. Often times when one grabbed a dead Jap by the arm or leg, it would pull loose from the body . . . or his stomach would burst open and his guts and maggots would spill out. Some of our guys had to vomit before they got about six bodies in the hole." Scheer did not vomit but he never forgot the experience.[20]

Souvenir hunters were having a field day, leading Baglien to comment in his diary that almost "every man in Company L has a Jap sword or pistol." Bill Welander was part of a team who searched the dead for maps, documents, and other useful information. "That was a horrible job," he recalled. "The bodies were swelling up under the sun and the smell was terrible." Later, Welander joined a patrol that followed the trails the Japanese had cut through the jungle during their march. Along the way, the patrol discovered the bodies of about thirty Japanese who were part of a headquarters unit. "They had all killed themselves. About half had their throats cut and the rest had committed hari kari." The casualty ratio from the battle was incredible. Against almost 3000 Japanese, losses in the 164th came to thirty-five dead and "about fifty" wounded. Most of these casualties were in the sections where "hand to hand fighting was common."[21]

Journalists and historians later called the carnage of October 24-27 the "battle for Henderson field." But to the men who were there, it is best remembered as the battle that gave the regiment its reputation as a fighting outfit. "The boys are hardened warriors," Baglien boasted. "Boy, how my regiment can scrap." Chesty Puller, who could seldom be made to utter a compliment about the Army, told Bryant Moore that his men were "almost as good as marines." Vandegrift sent Moore a formal commendation, expressing his pride to be serving with "another unit that has stood the test of battle and demonstrated an overwhelming superiority over the enemy." In Hawaii, Admiral Nimitz composed a compliment that expressed his vision of his entire Pacific strategy: "We feel that you [on Guadalcanal] have welded a combination that will be more than a match for the enemy."

But for the men of the 164th, the greatest compliment likely came from an unnamed First Division marine. Vandegrift was already bringing men from the 2nd Marine Regiment over to Guadalcanal from Tulagi. He needed them for his planned offensive across the Matanikau River. John Stannard and others heard one of the arriving marines make "disparaging remarks" about the Army. Immediately, one of the Old Breed went over and shoved the man, yelling, "You dumb bastard, that's the One-Six-Four. You don't talk to the One-Six-Four that way!"[22]

In turning back the Japanese thrust, the American troops had presented General Vandegrift with an opportunity to go over to the offensive. Events were moving quickly, both in the southwest Pacific and across all the other theaters of the war. During the second and third day of the struggle for Henderson, the U.S. Navy had blocked another attempt by the Japanese navy to neutralize Henderson Field with carrier air strikes. In what came to be called the "Battle of the Santa Cruz Islands," American dive bombers flying off the carriers *Enterprise* and *Hornet* damaged all four of the aircraft carriers in the Japanese task forces. Enemy counterattacks sank the *Hornet*, but the Japanese were forced to withdraw and cancel a large-scale reinforcement of their forces on Guadalcanal. Knowing this, Vandegrift decided that now, while Maruyama's forces were still retreating through the jungle, he could send his own men across the Matanikau, seize the high ground around Point Cruz and the push the Japanese heavy guns out of range of his airfield. He would use the 2nd Marines and some companies of the 164th's 1st Battalion to make the assault. On October 29, Halsey promised Vandegrift that he would get the 8th Marines and "an Army regiment" to the island as soon as shipping was available.

Shipping was the key factor in winning the struggle for the Solomons. Whoever could control the waters long enough to get their reinforcements into Guadalcanal would win. The U.S. barely had the upper hand at the start of November. The *SS Coolidge* was lost at the end of October, when it was carrying a regiment of the 45th Division into the harbor at Espiritu Santo and struck a mine. It went down quickly. Virtually all of the 5300 men abroad the ship survived, but their equipment was lost. In Washington, President Roosevelt was preoccupied with "Operation Torch," the Anglo-American plan to land some 125,000 men on the shores of French Algeria and Morocco in early November. Vandegrift could only dream of having such power in his hands. Roosevelt took time to approve more shipping for the Pacific and sent letters to the Joint Chiefs ordering them "to furnish immediately all assistance within [their] power to the defense of our position [on Guadalcanal]." The Navy complied with fast supply runs that were dropped off shore in floating drums for the men to pick up. The Army sent more anti-aircraft guns to the Pacific. Most important, the Army Air Force agreed to send some of its P-38 'Lightning' fighters to join its 67th Fighter Squadron at Henderson. Well armed and armored, capable of excellent performance at high altitude, the Lightning would make short work of Japanese bombers and could hold its own against the Zero fighter. When the P-38s arrived on November 12, "the Marines on the field spotted the Army pilots climbing out of their formidable fighters [and] they cheered wildly and rushed out to greet them." "This," Vandegrift, one of Vandegrift's staff later wrote, "broke the log jam and brought about the ultimate victory."[23]

It took time for all the assistance to arrive. For the time being the 164th remained the only large Army force on the island. Pleased as Vandegrift was with the soldiers' performance in defense, he was not yet certain of their abilities in the attack. So he planned to use his own 5th Marines and two battalions of the 2nd Marines in his opening thrust toward Point Cruz. The attack opened on November 1, with artillery pounding known and suspected Japanese positions, and naval guns suppressing the enemy artillery. Then the 5th Marines crossed the Matanikau and deployed two battalions. One battalion moved forward handily but the other, closer to the beach, was held up by machine guns that stubbornly conducted a "delaying action" and inflicted several casualties. Still, the marines made progress, advancing about a thousand yards on the first day. During the second day, they managed to push past Point Cruz, trapping a large number of the enemy in that pocket of beach and jungle. But the marines were taking casualties, and after three months on the island, their ranks, about fifteen percent of the First division's men, were bedridden with the illness. In order to wipe out the Point Cruz pocket and still advance, Vandegrift called for the 1st Battalion of the 164th to enter the fray and push on to Kokumbona. So the battalion was called forward and attached to the 2nd Marines.[24]

Bill Tucker and his machine gun crew from Company M were seconded to a rifle platoon for the advance. As they went forward on November 3, they were loaded down with extra rations and ammunition. "You couldn't run any vehicles up there so each one of us carried two mortar shells, and that was dangerous." Once in place, he began to provide cover fire for the attacking riflemen. "We had to be careful, because our water [jackets] would get too hot if we fired too many rounds at once. When we fired tracers, we worried that the barrels would overheat." The Japanese returned fire with their machine guns and with mortar rounds. "Those knee mortars, I'll tell you they were vicious, they could put them right into your hip pocket." The Japanese resisted tenaciously, putting all service personnel, even debilitated and wounded men into the fight. If a man could pull a trigger he took his place in the defenses.[25]

While under fire, Tucker remembered how magazine articles back in the States had assured American readers that the Japanese had poor eyesight. "Sure some had thick lenses on their glasses, but oh, those snipers could see real well." A few days into the battle, Tucker and his friend Sergeant Bob Bjerke, crawled out to help another man who had been wounded in the knee. They took him back to the command post where a medic gave him a "little bottle of brandy." The man smiled and said "Doc, why don't you shoot me in the other knee and give me another bottle."[26]

The attack continued and now the 1st Battalion began to take serious casualties. Men who had no combat experience in offensive maneuvers too often moved too close together. And modern automatic weapons rarely give exposed men a second chance. On November 5, Company B pressed forward. One of their platoons moved into a well prepared Japanese crossfire. Walt Byers, a rifleman with Company B, watched the platoon as it advanced. "They were half way down this grassy knoll [on the reverse side the hill] and were headed toward this little valley. The Japs had deployed [their snipers and machine guns] in a horseshoe formation." When the platoon was well into the trap, the guns opened fire. "We lost practically all those men. The wounded were calling to us to come get them. It was terrible. We had to put some artillery fire down on the [enemy] positions and shells landed among those men."

Les Wichmann also witnessed the tragedy: "They were caught in an entrapment with enemy machine guns on their left and right flanks, and one in front. Pvt. Sherman Olson, Pvt. Gerhard Mokros, Pvt. Stanley Ziska, Sgt. Herbert Langard, Sgt. Robert Cross, and Sgt. Raymond Johnson were killed." Men who tried to go forward and bring the wounded back were hit themselves. "Lt James McCreary and S. Sgt Arthur Jones went out to try and assist the platoon and they were also killed. A medic named Pvt. Carl Hjelm was also killed. Even now, it hurts me to write about this." One of the men caught in the firing dove into tall grass and came across Private Sherman Olson, who was badly wounded and helped him with morphine. Lying in the grass alongside Olson, he could feel "the bullets flying less than a foot over [my] head and grass was falling all over." Hearing a noise he looked up to see a Japanese soldier preparing to bayonet Olson. "I rolled over on to my left side, lifted my rifle and fired. My round hit the standing [Japanese] soldier in the area below his left arm pit lifting him off the ground and tumbling backwards." The man was able to crawl back to the American positions. Lloyd Sweeney was also caught in the crossfire and lay wounded on the field, "playing dead" for over a day before he was able to crawl back. By the end of the day, the First Battalion had a "definite shortage of officers" to continue leading the attacks.[27]

The advance was costly but the men were making progress when a fly leaped into the ointment: a captured Japanese document revealed "that Admiral Yamamoto and General Hyakutake [commander of the Japanese 17th Army units in the Solomons region] would land advance elements of the 38th Division east of Koli Point beginning at midnight" November 3. Koli Point was about five miles east of the American perimeter. In the words of Merrill Twining, one Vandegrift's staff officers, "we did not want Japanese astride [us] on both east and west flanks." So Twining and others persuaded a reluctant Vandegrift to slow up the Matanikau advance, send one battalion of the 7th Marines by sea, and a large part of the 164th on foot, to Koli Point. What the Americans did not know was that the Japanese, hurting now for shipping, had already cancelled the Koli troop landing. Now only a supply run was scheduled to bring food to the retreating force of Colonel Shoji. The "fog of war" was leading the Americans into a battle that may have delayed the victory on Guadalcanal.28

On November 4, the 2nd Battalion of the 7th Marines marched out toward Koli Point. Just west of Koli they collided with Japanese troops in the dense jungle. Most of these Japanese were part of the resupply effort that had landed to meet Shoji's ragged survivors, but they were armed with heavy mortars, that inflicted casualties on the marines. Communications back to Vandegrift and Twining were poor due to wet radios and the confusion was multiplied when U.S. planes bombed the marines instead of the enemy. The marines fell back and dug in. Vandegrift ordered Moore to take his 2nd and 3rd battalions to Koli Point to hit the Japanese flank. Vandegrift placed Brigadier General William Rupertus, his assistant division commander in charge of all three battalions. No one had any clear idea where the Japanese were in the jungle.[29]

Bryant Moore accompanied his two battalions on the march (see Map 3). Also along was General Edwin Sebree, the assistant commander of the Americal Division, who had come to Guadalcanal to prepare for transferring the rest of the Americal to the island. Whether or not having two generals looking over his shoulder while his men moved out made Moore uneasy is not known. What is known is that Moore was personally in a very challenging spot. As an Army colonel leading an Army unit under a Marine command, a failure could readily ruin his career. He was a very capable man, methodical and intelligent and a good organizer. But while few of the rank and file harbored anything against him, there were many in his own headquarters who resented him for replacing Colonel Sarles. He was held in suspicion by his battalion commanders. His assistant commander, Baglien, was openly hostile toward him.[30]

Moore in turn found deficiencies in several of the company officers. He doubted their knowledge of modern weapons and tactics and thought they were ignorant about proper staff procedures. Above all, he thought too many of the guard officers were "too chummy" with the enlisted men and were markedly reluctant to exert discipline. He was already making up a list of those he felt should be replaced. Koli Point would bring Moore's differences with his officers out into the open.

(Endnotes)

[1] Vandegrift, *Once a Marine,* pp. 186-87.

[2] Frank, *Guadalcanal,* pp. 357-61, describes the air and sea actions, while Merillat, *Guadalcanal Remembered,* pp. 200-207, has his original notes of actions reported to headquarters, which contains the "lie low" advisement.

[3] William Welander interview. The S2 Journal entries for October 25-26 are largely devoted to comments about the naval and aerial bombardments and reports of Japanese snipers. Nothing was recorded in the journal from 10:00 p.m. on the 25th until 6:15 a.m. (local) on the 26th.

[4] Stannard, *The Battle of Coffin Corner,* pp. 69-70. Japanese machine gun crews routinely dug their weapons down into the roots of enormous banyan trees. Before the employment of flamethrowers, knocking out such emplacements was extremely difficult.

[5] John Kasberger, telephone interview with Shoptaugh, August 30, 2007. Richard Stevens, telephone interview with Shoptaugh, August 29, 2007.

[6] Albert Wiest interview; John Kasberger interview.

[7] Frank, p. 362, identifies two of the 37mm guns at the junction of L and E Companies as belonging to the 7th Marines.

[8] Tollefsrud interview with Shoptaugh.

[9] Frank, pp. 361-65, provides with Japanese sources a detailed description of the enemy dispositions and how the attack occurred across the entire southern perimeter. See also the 164th regimental summary, "Report of Battle for Henderson Field," June 28, 1943, in 164th Infantry Association Records (the original is in National Archives). American regimental organizations did not provide for any designated J companies, to avoid any confusion in field orders with I companies. Thus, I, K and L made up the rifle companies in Hall's battalion.

[10] Tollefsrud interview with Shoptaugh, and James Fenelon interview of Alvin Tollefsrud, July 6, 2000 (copy provided courtesy of Fenelon).

[11] Stannard, *The Battle of Coffin Corner,* pp. 70, 78; Walker, *Combat Officer,* p. 23; Frank, p. 362.

[12] Neal Emery's memories from an interview by W. Michael Morrissey, June 14, 2005, deposited at the Barnes County Historical Museum, Valley City. Rolland Dahl's actions were first reported by Foster Hailey of the *New York Times,* then reprinted as "He Killed 100 Japs," Valley City *Times Record,* June 5, 1943. The article places the events Dahl described as happening on the night of October 23-24, which is not correct. Dahl mentions for example that Lt. Robert Kammen died in action the same night which was October 25-26.

[13] Don Hoffman's remarks from interview with Shoptaugh. John Hagen's taped memories are related in more detail in Shoptaugh, "Never Raised to be a Soldier: John Hagen's Memoir of Service with the 164th Infantry," *North Dakota History,* Winter and Spring issues, 1999. See also Wiens, "My Own Little Corner of the War," p. 95.

[14] Stevens telephone interview with Shoptaugh.

[15] A fine account of the Marine Corps portion of the October 25-26 battles are in Hoffman, *Chesty,* pp. 192-93, and Hough, *Pearl Harbor to Guadalcanal,* pp. 336-37. See also Stannard, pp. 88-89. The P400 Airacobra was a version of the P39, a fighter plane that was well armed for attacking ground targets but lacked the altitude and power to successfully engage Japanese Zeroes.

[16] Maruyama's remarks about the "Decisive Battle" and lacking food or supplies after the battle are in Samuel B. Griffin, *The Battle for Guadalcanal* (Philadelphia: Lippencott, 1963) pp. 203-04, and in Frank, pp. 363-64. See also Walker's comments in *Combat Officer,* p. 24. Later in the war, Chesty Puller talked to one of the rare Japanese prisoners and asked him why his officers never would not adjust their tactics after head on attacks at entrenched positions proved futile. The man replied "The plan had been made. No one would have dared to change it. It must go as it is written." See Burke Davis, *Marine: The Life of Chesty Puller* (New York: Bantam Books, 1988) p. 144.

[17] Wiens, pp. 92-93; Tollefsrud interview with Shoptaugh.

[18] Stannard, p. 80; Tollefsrud interview; Baglien diary entry for October 27, 1942; John Hagen tapes at NMHC.

[19] Paulson interview with Shoptaugh; Lauter statement in "One Man's Story," *164th Infantry News,* March 2006.

[20] Baglien's article of "The Second Battle of Henderson Field," relates how Morgan won the Silver Star. Bernard Scheer's account from his "Resume of My Army Career," dated December 7, 1945, copy given to Shoptaugh by Daniel Scheer, Bernard's son.

[21] William Welander, interview with Shoptaugh. Baglien's diary lists the regiment's "total loss" as of October 29 at "35 killed and 54 wounded." Japanese casualties from Frank, p. 364.

[22] The official commendations and citations for the 164th are reproduced in Vandegrift, Frank, and Miller; Stannard, p. 114, for the "One-Six-Four" incident. Baglien's words of pride in his regiment are in the transcript of his diary. According to Harry Wiens, Puller told Hall: "God, Bob, you should have been a Marine!" -- *My Own Little Corner of the War,* p. 121.

[23] Miller, p. 172; Center for Air Force History, *Pacific Counterblow: The 11th Bombardment Group and the 67th Fighter Squadron in the Battle for Guadalcanal* (Washington DC: United States Air Force, 1992 reprint of 1943 ed.), pp. 45-

56; Merrill Twining, *No Bended Knee: The Battle for Guadalcanal,* (Novato, CA: Presidio Press, 2004), p. 174.

[24] Miller, pp. 190-95.

25 A 'knee mortar" was actually a Japanese grenade launcher that the enemy usually fired from a crouching position. Because the butt plate of ths weapon was curved , American soldiers thought the Japanese soldiers fired the weapon with the plate over their knees. It fact, such an act would likely break a man's kneecap. As Al Wiest noted, "it kicked like the devil, the butt plate was placed against the ground."

[26] WilliamTucker interview with Shoptaugh.

[27] Walt Byers interview at Camp Grafton reunion (undated) in North Dakota National Guard History Files, Bismarck; Les Wichmann e-mail to Shoptaugh, March 22, 2008; Allen J. Olson, "Experiences of Sherman Olson on Guadalcanal" (1985), file in 164th Infantry Association Records. The officer shortage is noted in the S1 Journal entry for November 5, 1942.

[28] Vandegrift, pp. 193-95.

[29] Miller, pp. 195-97.

[30] Interviews with several men of the regiment reveal the level of anger toward Moore. Douglas Burtell was a seventeen year old mapmaker and reconnaissance rifleman attached to the HQ, who vividly related the contempt for Moore in his October 2007 interview: "He'd sit down in his shelter and gripe 'They're not fighting hard enough! Nobody was killed today!' Can you imagine hearing that?" James Fenelon, a staff sergeant in the HQ Company, James Beaton, a young officer in the S2 (intelligence) section, and Melvin Bork, a clerk at the 3rd Battalion HQ, all expressed an extreme dislike of their new commanding officer in interviews with Shoptaugh in 2007.

Chapter 7: Feeding the Mamolo
Guadalcanal: November 1942

In 1896, a group of German explorers were murdered on Guadalcanal because they climbed Mount Tatuve, the highest peak on the island. Ascending Tatuve offended the islanders who regarded the height as the sacred home of "the Mamolo," a godlike creature who demanded blood sacrifice and was said to eat human flesh. So the natives killed the Germans and westerners avoided the island for years. Then the lure of palm oil and some gold deposits drew the westerners back. The Australians ultimately took control of Guadalcanal until the Japanese seized it after Pearl Harbor. The Americans then came to take it away from the Japanese and by November 1942 there was enough blood flowing to feed a hundred Mamolos.[1]

By November 5, the 2nd Marines and the 1st Battalion (164th) had succeeded in driving a wedge into the Japanese positions west of the Matanikau River, trapping and killing over three hundred of the enemy in the Point Cruz pocket. William H. Whyte, a marine intelligence officer, noted that the pocket "was honeycombed with machine gun positions . . . skillfully camouflaged with leaves and branches." Whyte believed that the Japanese in the pocket were cracking under the attacks but expected that the battle to finish them off would be protracted and costly. He was right. Lieutenant Carl Vettel led a squad from the 164th's Company A against enemy positions in there a few days later and was killed while throwing a grenade at a machine gun nest. Several others were killed, too.[2]

General Vandegrift studied his situation maps and was satisfied that the pocket would fall and his attack toward Kokumbona would continue to make progress. If he could concentrate most of his units in that direction he thought he could drive all the way through the Japanese defenses and end the campaign. But he was worried that the Koli Point situation might interfere with the advance. If the enemy did land major forces there, as the captured enemy documents suggested, then he could once again be forced to deal with threats from two sides. Vandegrift decided that he had to temporarily halt the westward thrust against Kokumbona and deal with Koli Point. He urged Bill Rupertus to push forward quickly, drive on Koli Point and then to Tetere, and destroy the enemy. Get it done fast, he ordered, so he could bring the soldiers back and resume the attack in the west.

Vandegrift's intelligence, unaware that the Japanese had cancelled the plan to land major forces at Koli Point, had overrated the eastern threat. Several hundred Japanese had landed at Koli, some infantry with supply platoons that were to meet with Shoji's men as they retreated from their drubbing in late October. Shoji had somewhere around 2000 men. Many of them were wounded, all were hungry, and they had almost no heavy weapons. They were in no condition to fight a pitched battle against the preponderance of Americans closing in on them. Vandegrift increased this force even more by getting permission to order the 2nd Marine Raider Battalion, just landed at Aola Bay even further east along the coast, to come west and attack any Japanese they met.[3]

The 2nd and 3rd Battalions began their march to Koli on November 4 (see Map 4). Patrols guided by islanders went ahead to reconnoiter the Nalimbiu River, which emptied into the sea at Koli. One patrol spotted a couple of small boats on the Nalimbiu and recommended to headquarters that they be attacked by aircraft and destroyed. Another patrol captured one Japanese prisoner, probably a straggler from Shoji's moving column. Meanwhile, a patrol made up of Bob Dodd, John Slingsby, and Doug Burtell, all part of the S2 (intelligence) section, went out further east and scouted the Metapona River. There, Burtell recalled, they found "a whole slug of Japs coming across the river." This may have been some of Shoji's survivors. The three men hurried back to report. The jungle was very thick in this part of the island. Sam Baglien wrote later that the "terrain is dense, steam-heated jungle, and progress is slow, requiring tremendous effort of all men. Necessary to cut [marching] lanes in many places."[4]

Japanese snipers held up the march. They used very good smokeless powder which made them hard to spot. Artillery fire was directed at suspected enemy emplacements. The navy had promised to send several destroyers to bombard the shoreline after dark, but the amount of fire support was limited as the 164th was running short of 60mm mortar ammunition. Mortar shells were flown up from New Caledonia for the two battalions, but it was tough to get such bulky loads up to the gunners. The 2nd and 3rd battalions marched together on the 4th, but split into two columns as they approached the Nalimbiu. As was his custom, Colonel Hall was up at the front line as often as he was at his battalion's HQ. Melvin Bork acted as one of Hall's escorts during some of these "personal reconnaissance walks." He recalled that at one point Hall, out ahead of his bodyguards, came upon a Japanese sniper who had been flushed out of his nest and shot. He was wounded but still alive. As the Japanese reached for a knife, Hall pull out his .45 revolver and shot him square in the chest. "Search him for papers and catch up with me," he called to Bork, and kept going up the trail.[5]

The march was hampered by the complete lack of any accurate maps. "Our maps at that time were so poor we might as well have used a map of North Dakota," Jim Fenelon recalled. None of the sketch maps given to the companies contained accurate locations of the rivers or other prominent features. The marchers were literally groping their way through the jungle. Chuck Walker provided a vivid description in his memoir. "Word came that a Japanese battalion had landed in the Tetere area and we were to assist in rounding them up. The next morning [November 4] we went cross-country through seven-to-eight-foot-high kunai grass. There was no breeze and the sun was hot; we sweltered in the humid heat." Each man carried two canteens of water, but it was very humid and they used it up quickly. It was tough to see much beyond the sweat-stained shirt of the man in front.

As darkness came the men could feel lighter air coming from the ocean and hoped that they were getting close to the beach. Suddenly, Walker thought he heard some low voices coming from the jungle. He called over to John Gossett, his commander and friend, and said he thought there were Japanese somewhere close. Another man agreed with Walker, saying "I can smell them." After a brief look around, the men found nothing. Gossett posted guards and everyone else settled down to sleep.

At dawn they began to eat cans of C rations, but tossed them aside when machine gun fire broke out in the jungle. A Japanese platoon had deployed in the night and now took part of Company F under fire. The GIs returned fire while Walker and a machine gun crew joined riflemen in Company E and set out to try and flank the Japanese. But they ran into more snipers as they worked their way into the heavy undergrowth. A man was suddenly hit and wounded by a sniper. Gossett looked over to Sergeant Miles Shelley, moving in with his Company F platoon, and pointed to a tree. "Are you sure," Shelly called back. Yes, Gossett nodded. Shelley worked his way to the tree while the others gave cover fire, and then, as Walker watched, "moved around its base in a tight circle, then he aimed up and fired." The sniper fell to the ground, "almost in slow motion."[6]

Over in the ranks of Company G, there was another fire fight. There, one of the lieutenants was hit in the stomach by a machine gun burst. Doc Schatz, the battalion surgeon, dashed forward and pulled the wounded man back to safety, then treated his wounds. Schatz won the Silver Star for this feat. The firing went on until mortar rounds forced the Japanese back to the north. The corpses they left behind were dressed in clean uniforms and so were probably part of the supply force that had landed on the beach. Both the 2nd and 3rd Battalion resumed the march to the Nalimbiu, crossed it and pressed on in two columns toward Koli. Sam Baglien was marching along with Art Timboe and the 2nd Battalion headquarters men, when he learned that his old friend from Hillsboro, Sergeant Albert 'Twinks' Osman, had been killed that day. "He was hit in the head. Three

Nip mgs and snipers held up our advance." A few hours later, American planes mistakenly dropped bombs near the 2nd Battalion's command post and destroyed some radio equipment. Baglien was becoming upset about the regiment's casualty list, which he put at "42 KIA, 9 MIA, 111 WIA" to date.

Back at regimental headquarters General Sebree was getting annoyed by the delays in the advance. He feared the enemy might slip away before Moore's battalions got into position. Like Vandegrift, he wanted Koli Point cleaned out and the troops brought back to go after the Matanikau hills. Moore noted that the 2nd Battalion companies had been delayed by the snipers, allowing the 3rd Battalion to get well ahead of them. Wanting his soldiers moving faster, Moore ordered the 2nd battalion to "push north to the sea tonight."[7]

Night marches were generally avoided because of the danger that friendly units would stumble into one another or into an enemy trap. But now the 2nd Battalion kept going. Company F was in the lead position for the 2nd battalion and some of the men were worried that the pace was too slow. Rudy Edwardson told his officers that the men were being allowed too many breaks. If they approached the 3rd battalion lines after nightfall, he feared there could be trouble. But the clashes with the small Japanese groups had made the men wary and they did not want to rush into an ambush. "We took too long to get there."[8]

It was very dark by the time F Company's lead men came up to the 3rd Battalion's night perimeter. It was there that the tragedy occurred. Lieutenant Walker recalled: "After crossing open field, we entered a narrow stretch of jungle. There a sudden burst of machine gun fire ripped into our column from the north. Then a second gun opened up. One of my section sergeants cried out, 'Those are Brownings firing – our guns!'" The 3rd Battalion guards had heard the marching men coming closer, assumed they were Japanese, and opened fire. The machine gun fire tore into the lead platoon in F Company. Phil Engstrom was one of the first men hit: "About 3:00 a.m., as we were moving along in pitch darkness, all hell broke loose. Machine gun fire and rifle fire were spraying bullets all around us. I was hit in my right knee and it spun me around. I fell back on to the guy behind me. I just can't remember his name, anyway he was dead, and I was trying like hell to get off him! We had run into our own 3rd Battalion men and they thought we were Japs, so they opened fire on us. It was a mix up in communications and a hell of a mess." Engstrom was wounded in three places. He was given some morphine by Harry Overbeck, one his childhood pals from Glenfield, who also splinted Engstrom's leg with his own M1 rifle. Then Phil was put on a blanket and carried to an aid station. "Needless to say it was a bumpy ride. My butt was hitting the ground every step of the way." Going back with him was Walt Rivinius, who "was hit by a bullet [that] splintered my whole ankle." In all, about a dozen men in Company F had been hit by friendly fire. At least two had been killed.[9]

The men were shocked by this incident. Many later blamed Moore for ordering the night march. Walker wrote, "this order to move had been made by our regimental commander who should have known better; the short move [at night] was absolutely unnecessary." Back in the headquarters, Jim Fenelon saw Moore's reaction to the incident: "When we got news of it, Moore said 'that was my mistake,' and I wish we had used that to get *him* relieved. But we were all young and didn't know all that much. But it was Moore's fault."[10]

Shaken, the two army battalions set off again in the morning to close the net on the Japanese. General Rupertus, shaking with dengue fever, complained to Moore that the soldiers were still not moving fast enough. The thick jungle and lack of water had the medics dealing with large numbers of men felled by dehydration and the heat. Men were forced to dig into the gritty soil to find brackish water that they purified with iodine tablets. Some inevitably became ill with dysentery. They reached the beach on November 6, making contact with some men from the 7th Marines. The Japanese were already withdrawing to the east. Rupertus ordered his makeshift regiment to press on toward the Metapona River. The estimate of the enemy forces had by then been reduced to about two weakened battalions. The Japanese were in no shape for an extended fight. Shoji apparently ordered the healthiest men to hold back the Americans while the rest hoped to slip away.[11]

On November 8, the American forces crossed the Metapona and closed in on Gavaga Creek, where the Japanese defenders were dug in. The 1st Battalion of the 7th Marines took the left flank, running from the beach southward along the creek, while the 2nd Battalion of the 164th extended to their right along the bank of the creek. The 2nd Battalion then pushed beyond the creek to surround the enemy. Vandegrift was by now confident of success and he ordered the 3rd Battalion of the 164th to return into the perimeter. He wanted them in order to resume his Matanikau attack. Vandegrift's confidence notwithstanding, he lost the services of two of his own best officers on November 8. Chesty Puller, 7th Marines commander, was wounded by a shell burst. And Rupertus, his dengue fever having worsened, was evacuated. Sebree took overall command from Rupertus. Sebree still wanted to use both Army battalions to envelope the enemy forces, but Vandegrift's orders meant that Sebree would have to carry out his envelopment with fewer men.[12]

Gavaga Creek was practically a morass of swamp, densely packed with thick, mossy vegetation standing in filthy water. In this setting a sudden battle would flare up as the advancing soldiers made contact, then just as suddenly stop as the Japanese fell back and faded away into the brush. In one such fight, Neal Emery lost a member of his platoon, who was shot through the knees by a Japanese machine gunner. "I helped carry my man back to the aid station. He was out of the war." A similar fight centered around the man that John Stannard regarded as the "finest fighting man in the regiment." Joe Miller was one of those who had transferred into the regiment as it left the states for the Pacific. "Joe was a loner, sullen and silent, he kept to himself. He had a reputation with the NCOs as a shirker, a griper, and a general troublemaker." But once on Guadalcanal, Stannard wrote later, "the biggest eightball in the outfit [became] the best man of us all." At Coffin Corner, Miller held his foxhole like a born warrior, using his rifle and grenades to drive off one attack after another, killing over thirty of the enemy. "There was never any doubt about Joe after that night." At Koli, Miller and Stannard were in a patrol that ran into an ambush. As a machine gun fired on them from the front, Japanese riflemen attacked from the flanks. Miller led the patrol in a counterattack. "He was everywhere, protecting the wounded, throwing grenades, and firing his M1 in clip-long bursts." The Japanese withdrew, leaving behind several dead. The patrol lost one man and had five wounded, which Miller helped carry to the rear. He himself had almost been killed when a bullet went "through his helmet and had creased the top of his head." Miller was promoted to squad sergeant the next day. He would be in charge of his squad for just four days.[13]

The 2nd Battalion flanking force was led by F and E companies that were directed to close the jaws on the enemy south of Gavaga Creek. At one point they walked across sandbars and shallow water to get into position. The area around the creek was poor ground and it was tough work cutting a way through the kunai grass with knives and bayonets, the men slogging in muddy, ooze-filled swamp. Japanese machine gun nests slowed the advance, and the enemy mounted small scale counterattacks with small arms and grenades. Jim Campbell, a private from Melville, North Dakota, faced one these counterattacks with his squad. A fairly big man, Campbell used his BAR to drive the Japanese back. His gun jammed no less than three times, but he cleared it each time and kept spraying the enemy until his squad was safe. Only after it was over did he realize that a bullet had taken off the tip of his index finger.[14]

Sergeant John Paulson's squad led Company F as it sought to find the main body of the Japanese. "My company commander, Captain Barker, said 'John, take your men and go straight toward the ocean. We'll follow behind you. When you make contact with the Japs, place your men and guide in the rest as we get there." It was a hot day and Paulson was not happy with the plan. "We were chopping through the grass, which made a lot of noise. So I said to Bob Cole, 'we'll be ambushed this way.' I thought we should move off and find a path and follow it. I decided we should go northeast and see if we could do that. After about twenty minutes we did find a trail, a nice path, no tress or bushes, and I said 'let's take that.' Maybe that was luck, but I still think seeing that path was something of the power of God.

The path went on until we reached some rough, trashy scrub and as we moved around it, we were face to face with five Japs coming toward us [carrying] a heavy [machine] gun. They didn't know we were there." The Japanese had not expected to run into any Americans on this trail and did not react quickly enough to return fire. Paulson's men quickly killed them, then advanced further down the trail and carefully scouted about, finding a "large number" of Japanese in a deep ravine. "I sent a man back to Captain Barker to come up and we got the company from that point into a line going east to E Company. We went forward a bit and pretty soon heard some voices in the tall grass. We took cover and I called out, 'come out.' A voice called back 'we malines.' They couldn't say that r sound. Well, that did it. When they wouldn't come out we started firing."

Company F then called for artillery fire. Once the shells began to land "all heck broke loose" around the American positions. Paulson "was by a big tree, standing there and suddenly a [Japanese] machine gun opened up. I dove behind the tree but then heard the bullets coming through the tree right above my head. I said 'Lord I don't want to die here, but it's up to You,' and I felt calmer." For several hours, the American soldiers poured fired down into the ravine, supported by mortar and artillery fire, including rounds from 155mm guns that had been delivered to the island just a few days before. "The next morning, we fixed bayonets and went in there – there were over three hundred dead in there."[15]

As Paulson's company finished off this group, other Japanese units were trying to slip south across the swamp. Company G was sent into action to close the gap between companies E and F. "But they were on the other side of the creek," Paulson noted. Men fought a series of small actions, often in hip deep muddy water mixed with blood. Chuck Walker's machine gunners joined Company G riflemen to tighten the ring around the Japanese. As they closed in, Walker "went forward to the Company F command post for information. There I found our Company H mortar officer, Lieutenant Clark . . . Just then a mortar round burst inside the Fox command post, wounding several men. I lucked out, but Clark's wounds were fatal." By examining the mortar fragments later, it was determined that the shell was an American short round. Company G pushed into the swamp all day, but the gap was never completely closed due to the poor ground and the fact that most platoon and squad leaders had not yet acquired the experience necessary for that kind of inter-unit coordination.[16]

Over in E Company, Corporal Joe Miller and his rifle squad moved along a native trail, looking for a fight. When they came into a small clearing, they were attacked by "a platoon of Nips [who] appeared on the other side of the opening." As the Japanese charged, firing from the hip, Miller returned fire until he was hit. He told the rest of the squad to "get back to the company" while he held the enemy off. They heard him fire off several more clips, but knew he probably wouldn't get out. The next day, Sergeant John Stannard took another patrol to the clearing "where Joe had made his stand." There, they found eight graves, one of them Miller's. "They [the Japanese] might have buried him out of respect for a real fighting man – I like to think that was their reason."[17]

On November 12, when the last enemy soldiers were wiped out, the total Japanese loss in the pocket was estimated at about 500 men. Marine and GI casualties came to forty killed and about 120 wounded. Sebree wanted Timboe to sack some company commanders for failing to close the trap. But Art Timboe refused to agree to this, arguing that he had good commanders and men who had "killed scores of Japs." Moore then relieved Timboe of command of the 2nd Battalion and Major Ben Northridge took over the unit. Failure to close the trap resulted from many factors. After the 3rd Battalion was withdrawn by Vandegrift, there were in several officers' views "too few men to complete the encirclement," pin the enemy, and then annihilate them. Memories of the friendly fire disaster snafus earlier perhaps played a part. Men were cautious about the prospect of running into their own guns. Chuck Walker remembered how he took care to call out his unit as his men reached the creek, lest the Americans he was coming up to "decide[d] we were Japanese."[18]

Vandegrift himself did not appear upset with the outcome of the battle. After the war, he wrote that the "combat inexperience" of the 164th had allowed some of the enemy to slip the noose, but that he "was not particularly surprised" by the result of the offensive, because the attack plan was "hastily improvised." As for his own decision to withdraw Hall's 3rd Battalion, which could have closed a tight ring around the Japanese, he decided that resuming the advance in the west was more important than trapping every enemy soldier in the east. In the next few weeks, the 2nd Marine Raider Battalion tracked the Japanese who had fled and killed as many as five hundred more men. Scores more died of wounds or fever in the jungle.[19]

Alexander Patch, the Americal Division commander, gave Timboe a new assignment in the Americal Division headquarters, and then took him along as part of his staff when he (Patch) was made commander the XIV Corps. Patch praised Timboe's work for the remainder of the latter's assignments in the Pacific. But Moore's relief of Timboe infuriated many men in the regiment, especially those in Timboe's battalion. Sam Baglien, many of those at the regimental headquarters and a number of those who had been in the National Guard before 1940 spoke of Moore with contempt thereafter. Others decided that this was a fight among the 'brass' and none of their business. Meanwhile the war went on.[20]

Koli Point had achieved the goal of eliminating Japanese threats from the east. But the time and effort taken at Koli had given the Japanese a chance to shore up their defenses in the Matanikau hills. In 1946, Japanese commanders who had survived the Guadalcanal campaign told American interrogators that in early

November they had been on the verge of a complete collapse in front of Kokumbona. Then, when Vandegrift had to deal with Koli Point, the Japanese gained time to reorganize. Parts of the Japanese 38th Division were landed at Kokumbona by nighttime destroyer runs. They took up defensive positions, while a reserve force was positioned to counterattack the Americans.

On November 10, the Second Marines and the 1st Battalion of the 164th moved forward with Kokumbona as their objective. Using mortars to repress enemy fire, the riflemen set out to reduce strong points. The Second Marines, supported by heavy weapons from the 164th's First Battalion, pressed against the hills guarding Kokumbona. They met heavy resistance and casualties again mounted. The next day Vandegrift's headquarters received word from New Caledonia that another naval battle was shaping up for control of the seas around Guadalcanal. Vandegrift ordered the attacks halted. The forces withdrew to wait on the outcome of the navy engagements.[21]

In order to get the bulk of its 38th Division to Kokumbona, the Japanese Navy committed two battleships and a host of cruisers and destroyers to cover the transports. The battleships would also bombard Henderson Field. The U.S. Navy, warned by decoded intercepts, countered with the *Enterprise* and two battleships, some cruisers and destroyers. Halsey had made good on his promise to give Vandegrift all the support he had at his disposal. He ordered his fleet to get additional ground reinforcements to Guadalcanal and then engage the Japanese. Enemy air strikes against Henderson Field were meeting the P-38 fighters for the first time and the P-38s quickly proved to be more than a match for the Japanese Zeros. Sam Baglien was returning from Koli Point by sea and had a good view of the air battles on November 12: "Got caught in a Higgins boat, right in the middle of the bombing and ack ack from the ships. Saw three Nip planes go down." A first contingent of reinforcements arrived: "twenty-one officers and 116 enlisted men" from the 164th -- men who had been left behind in October and the reconnaissance platoon, commanded by Lieutenant Fred Flo. Additional reinforcements included two battalions of the 182nd Infantry Regiment (Americal Division), marine aviators and replacements, some engineers, and more artillery. In all, a most welcome increase in combat force.[22]

That night, the Japanese battleship *Hiei* with cruisers and destroyers, sailed in to shell Henderson. American cruisers and destroyers met this force head on. Soon after midnight the men on the island heard the thunder of big guns as the rival forces engaged in a fierce clash out in the channel. Both sides lost heavily: two American destroyers and one cruiser sunk and several other ships badly damaged, while the *Hiei* was so badly mauled by the American cruisers that American aircraft sank it the next day. Throughout the day bodies of dead seamen drifted in on the shores of Guadalcanal while American bomber pilots located the Japanese transports and hammered them, sinking six of the ten transports. Very few men of the 38th Division were going to arrive intact.

The climax of the battle came on the night November 14-15, when the second Japanese battleship, the *Kirishima,* supported by cruisers and destroyers, tried to penetrate the American navy's screen and attack the airfield. The battleships *South Dakota* and *Washington*, sent by Halsey to take station off Guadalcanal, met the enemy and blasted the *Kirishima* into a flaming, sinking wreck. The four remaining Japanese transports managed to get into the Sound during the melee, but as American planes attacked them at dawn on the 15th, all their commanders could do was beach the ships at Kokumbona and order the infantrymen to run ashore through a gauntlet of bombs and artillery fire. Hundreds died in the attempt. In all, only about a fifth of the ten thousand men who had set out to reach the island lived to get there. Virtually all the supplies, ammunition and food sent with the soldiers were destroyed.

Both sides had suffered grievous losses. The Japanese lost thousands of soldiers, sailors and air men. The Imperial Navy commanders refused to send anything larger than a destroyer into Guadalcanal waters again. Nine American ships in all were sunk by Japanese shells and torpedoes. Some 1700 U.S. sailors were killed. The Navy kept the losses secret for months. Richard Stevens noted that it was not until years after the war that he read about the extent of the naval casualties at Guadalcanal. "We had tough times on the island, but those young sailors really had terrible casualties," Stevens remarked. Vandegrift paid tribute to the U.S. Navy on November 16: "We believe the enemy has undoubtedly suffered a crushing defeat. . . . To [the Navy] the men of Cactus [Guadalcanal] lift their battered helmets in deepest admiration."[23]

With this defeat, the Japanese on Guadalcanal could do nothing but hold on as best they could. Vandegrift brought the 2nd Battalion back from Koli and sent the entire 164th into reserve while he prepared his next attack to take Kokumbona. The 164th was ready for a rest. "Everyone was so damn tired all the time, we were absolutely worn out," John Hagen recalled. Clement 'Murphy' Fox, a rifleman with Company D, agreed with Hagen. Travel writers might think "the South Sea islands are great," he wrote his sister Marie. But Guadalcanal was just "HELL." "Give me the Red River Valley in Dakota with all of its changeable weather for the rest of my natural life . . . if I do any kicking when I get back, I hope you kick my pants. I have remembered every prayer I have ever known."[24]

No one was anxious to storm the interlocking machine gun nests that guarded the approaches to Kokumbona. Told that Vandegrift had decided to commit the 182nd Infantry Regiment to its first action, they hoped that the "new guys" would carry the ball for a while. They were soon disappointed. With part of the 8th Marines attached, the 182nd's two battalions went into action on November 18. Toward evening, the untried men had their first casualties when an enemy ambush fired on men sent to get fresh water. The next day they moved west, covering the left flank of Company B of the 8th Marines, which seized a grassy knoll designated as Hill 78. Past the hill, machine gun fire forced the marines to halt. After sundown, Japanese mortars began dropping rounds and then near sunup the enemy struck with a quick, hard-hitting counterattack. Surprised, part of the 182nd broke and fled to the rear before regimental officers came up and restored order. Two companies of the 182nd then resumed the attack, but again it stalled.

By that time, the 182nd's men were "confused," and "some straggling" was occurring. Sebree, acting as tactical commander of the advance, decided that experienced men were needed to steady the newcomers. He ordered Moore to send up the 1st and 3rd Battalions of the 164th. Their brief rest was over. Watching the men march to their positions, dodging "fire by snipers," John Hagen commented that "there were some pretty pissed off North Dakota farmers up there, as mad at those Massachusetts boys as at the Japs."[25]

The reinforcements moved between the two battalions of the 182nd, forming a front about 700 yards long. The Japanese positions that the 164th's men faced were two of the main objectives, hills 82 and 83. The Japanese had honeycombed the reverse slopes of these with machine gun emplacements, many of them dug into the roots of large trees, covered by coconut logs and dirt, and then camouflaged by jungle vines. Even if the positions could be spotted, a direct artillery hit might not knock them out. To reach the hills 82 and 83, the men would have to first cross slightly lower hills, move down through a long ravine, and then cross some fifty yards of open ground before climbing the hill slopes, exposing themselves on the ridge lines. Only then could they take on the reverse slope emplacements. This amounted to a frontal attack, one in which the soldiers would have to fight for every foot of ground.

The Matanikau River was swollen with rain and the engineers had difficulty bridging it. Native labor hauled the supplies across the river by hand. Rain and mud made it all but impossible to cut vehicle roads into the hills, so no tanks could be brought up to help. It was a situation made to order for the Japanese defenders. Later in the war, the Army would use specially trained fire teams for these types of bunkers, men armed with explosives and flamethrowers, backed by support fire from tanks or anti-tank guns. Flamethrowers were being tested, but they were not ready. Pinpoint strikes from the air, or directed fire from navy guns, would require a better communications system. To take these hills the men had some artillery and machine gun support, grenades, and their personal weapons. Their only armor was their fatigue shirts and stout hearts.[26]

The attack began on November 21. P-39s and dive bombers from Henderson hit Hills 82 and 83. Then mortars and howitzers pasted what they hoped were "some Jap positions." Then at about noon the leading companies of the two battalions left their foxholes and set forth down into the ravine. Some enemy snipers fired on the men and had to be rooted out and killed. The only cover that the men could find as they advanced over the hills was some tall grass, which did little to hide them from the Japanese machine gunners. Men began to fall as the enemy raked the American soldiers with a torrent of bullets. Wounded cried out for medics, a buddy, anyone who could help them. Enemy mortars began to get the range and dropped lethal loads across the front. The attack stalled after advancing at most forty yards. "Movement very slow," read the regiment's daily journal entry. "Difficult to get ammunition up to lines. Casualties are heavy. Enemy appears to have set up strong defense. Occasional skirmishes and hand to hand fights. Efforts tax men to exhaustion." Sam Baglien's summary of the day's battle had a barb aimed at Moore: "The attack was stopped cold due to mutually supporting Jap mgs. Maybe the higher ups will realize you can't go into a stone wall."[27]

Whatever Baglien may have hoped, the 164th was ordered to attack again the next day. Again aircraft attacked the positions on Hills 82 and 83 and again the artillery delivered a barrage. But several hours passed before the men went forward, giving the Japanese time to recover their wits. Everyone was stunned by the manner in which the Japanese would stay at their posts and fight to the death. "They just accepted death," Dick Stevens noted. "I remember one night up there, a couple of Japs crawled up to our lines, killed two men, then tried to carry away one of our machine guns. They had no chance to get away before we killed them but they tried it anyway." The attack stalled again. Some enemy nests were destroyed by grenades, but at high cost. Moore was forced to order the men to fall back and "consolidate our positions on [the] ridges."[28]

Vandegrift was going to leave Guadalcanal with the worn down First Marine Division as soon as the last regiment of the Americal Division came up from New Caledonia. But he wanted to see the American forces established in Kokumbona before his departure. He decided that the 8th Marines should take a stab at the hills in front of the 164th. He ordered that after a brief barrage at 0600 to register the guns, the 1st Battalion of the 8th Marines would storm the hills. The GIs would remain in position, to exploit a breakthrough. Surely, Vandegrift thought, his marines would break through the enemy wall. But the heavy Japanese fire brought them to standstill, while the men of Hall's 3rd Battalion took new casualties from unfriendly artillery. It would take more than a frontal charge to dislodge the Japanese from these hills.

These three days of combat almost doubled the casualties of the regiment. Baglien tallied the losses to date as 117 killed, 208 wounded, and 325 men evacuated from wounds and fever. "We also have 300 men in the rear area with sickness, shell shock, heat exhaustion, minor wounds, and hysteria," he added. "Our combat strength is [now] less than 2000 men and we are facing a regiment of Nip Infantry with supporting weapons, well dug in."[29]

A lull settled over the battlefield as both sides paused to catch their breath. Dick Stevens' machine gun was in a long line of outposts that the men now called "the Limbo Line." "The U.S. occupied the high ground on one line of hills; the Japanese were on the other side, certainly out of hand grenade range. They were too weak to do anything aggressive. We didn't go there. They didn't come here." The men lived on cold rations, for a fire would draw mortar shells. At night, they could look out and see flashes from clashes between PT boats and Japanese destroyers trying to deliver supplies. Stevens remembered being happy at the sight of two large ships burning, assuming they were the enemy's. Only later did he learn they were American.[30]

Danger still hovered over the men. George Duis, a member of a machine gun team, awoke one night when a grenade exploded nearby. The enemy "had infiltrated our lines and attacked us from the rear." A second grenade sailed into the foxhole he shared with two other men. He grabbed it and threw it out but "exploded right away," wounding him and his mates. "Then the Nips hit us. I saw six of them [and] shot at them with my rifle and I'm sure I hit four. But they charged into the hole with bayonets, bayoneting me and killing my two unconscious pals. Duis pulled a pistol and shot the last attacker "while his bayonet was still in my stomach." Wounded by shrapnel and stabbed three times, fearing he would die, Duis remained awake until a medic treated him. He was evacuated home after emergency surgery.[31]

The frustration is still palpable in the personal memories of those who fought in this battle for the Matanikau hills. Les Aldrich, now a platoon sergeant in Company I, saw several of his men die in the first couple of days. "We had five guys killed when we made a charge [into the hills] and were driven back. We had to dig in and their bodies were out there for some days. Captain Knott said 'lets go get them the first cloudy night.' So we got up a group of volunteers. I went with Woody Keeble and two others along with Knott. We had a shelter half each and crawled up there and rolled the bodies on the shelter half." As Aldrich pulled the corpse of his man onto the canvas, the man's helmet rolled away. "The top of his head was shot off and I could see it was full of maggots. It was the worst thing I've ever seen."

Woody Keeble was a private in Company I, a very tall, muscular Sisseton Sioux Indian, who had been born in South Dakota, but grew up outside Wahpeton after his mother moved there to teach at the Indian School. Woody became known locally for his athleticism. He was a particularly good baseball pitcher, with a wicked fast ball. He played in every other sport as well. Many members of the Sisseton-Wahpeton Sioux, including Keeble, joined Company I at Wahpeton before America entered the war. Prejudice against Native Americans was prevalent in the region and the Sioux in Company I were not exempt from its stings. White men called them "chief" and made jokes about "going on the warpath." Almost none of the Sioux ever advanced beyond the rank of buck private. But they all fought bravely on Guadalcanal, none more so than Woody Keeble.

Because of his enormous size, he was made his platoon's BAR man, and he became proficient with the heavy weapon. To his comrades Keeble seemed completely fearless, but this wasn't true. He later admitted to feeling "the relentless and powerful hold" of fear, which he likened to a form of drunkenness that could rob a man of all reason. Pushing aside fear, he was always to be found in the thick of the battle, using his BAR to break up Japanese counterattacks, closing to near-suicidal range in order to assault machine gun nests with bursts of fire, using his pitcher's arm to toss grenades accurately. When Keeble's ammunition carrier was killed in the Matanikau fighting, he refused to take another man for the job and carried his BAR clips himself on his belt and in a shoulder harness; he did not want the responsibility of endangering another man. He threw himself into every battle in order to protect his mates. Jim Fenelon, who had known him since childhood, thought that in battle "the safest place to be was right next to Woody."[32]

But in the long run, no one was invulnerable. Bill Tucker had been at the Matanikau front since the beginning of November. On the 23rd, his luck ran out. "I was sitting talking to [Lieutenant] Al Whitney, from

Grafton, who I'd known since he was a sergeant. Our First Sergeant was also sitting there. We didn't have much of a foxhole, because you couldn't get up and dig with all the fire. The sergeant had just gotten hit in the hand, and I heard this shell come in. A big piece of metal went right through Al's chest. I got one [fragment] up here [above the lip], others in my right arm, left ankle and right knee. But I figure I was pretty lucky." Captain Andrew Panettiere, one of the medical officers, began to treat the wounded. "But then [the next day] Captain Panettiere got killed too."[33]

Bleeding from his face wound, Tucker crawled back toward a navy aid post. "They took me in and a man examined me and said 'we better get him out' so they covered me up with half a pup tent and that night they gave me a can of grapefruit juice but I couldn't sit up to even drink it. Then they flew me out on a DC-3. The only thing that bothered me was we were [flying] real close to the water to avoid any Zeros. Later I learned that the next morning Lieutenant Wiest wanted to check on me, but when he came, all he found was bloody tarp, so they didn't know if I had been evacuated or was dead. Nobody knew anything on the 'canal.'" Tucker was evacuated to Efate and after a few days was sent with several other wounded to New Zealand on a hospital ship.[34]

Even when under fire from the Japanese artillery, the men operating Henderson Field worked to get badly wounded men off the island for treatment. After they were wounded at Koli Point, Phil Engstrom and Walt Rivinius were given the finest care. Engstrom appreciated his company captain talking to him while he was treated at the main aid station. He "told me that I was thru with this war and that I would be going home soon. That was good news to me!" Rivinius, Engstrom and several others were flown out from Henderson Field. "They used a two-engine airplane and they set us [evacuees] under the wings, out of the sun, while they loaded up the plane with gasoline," Rivinius remembered. Then they flew us away, about six or eight of us. We got to a little hospital at some little island [Efate] and then eventually I ended up on Fiji." After treatment at Fiji, both Engstrom and Rivinius were returned to the states and spent several months at a new hospital in Tennessee, established for the rehabilitation of gunshot casualties. Both men were given medical discharges near the end of 1943.[35]

It was hard in a close-knit guard unit to see men they knew so well wounded and hauled away, their fates a mystery, sometimes for months or years. But the war had to continue and the losses accepted. The Japanese, themselves so afflicted with wounds and malnourished to the point that they called their posting "Starvation Island," still held on stubbornly, dying at their guns. All the soldiers on both sides could do was accept their fates, trading lives for yards. "We are killing and destroying Jap positions daily," Baglien wrote. "It is hard grubbing, but we will get them out."[36]

The bloodthirsty Mamolo had enjoyed a good November. No one else did.

(Endnotes)

[1] J. C. Grover, "Some Geographical Aspects of the British Solomon Islands in the Western Pacific," *The Geographical Journal*, (Sept., 1957), pp. 300-301. Dick Horton, *The Happy Isles* (London: Heinemann, 1965), p. 157, describes the legend of the Mamolo.

[2] Alvin Tollefsrud described the death of Carl Vettel in his interview with James Fenelon; William H. Whyte's patrol is noted in his book, *A Time of War*, pp. 55-62. Whyte later wrote a seminal study in human behavior, *The Organization Man* (1956).

[3] The best account of the Koli Point maneuvers is in Frank, *Guadalcanal*, pp. 414-23. The accounts by Miller, pp. 195-200, and Cronin, pp. 58-61, are good, but lack the Japanese sources in Frank. The 2nd Raider Battalion was at Aola Bay, along with other units, to build another airfield. This plan had been pushed by Admiral Turner for weeks, but the in the end the ground at Aola proved to be too swampy for viable air operations. Vandegrift had to get Halsey's permission to use the Raiders – there were too many complications in the American command structure at this point on Guadalcanal.

[4] Patrol information from Douglas Burtell, interview with Shoptaugh; Baglien diary, entry for November 4, 1942.

[5] S2 Journal entries for November 2 and 3, 1942, in 164th Infantry Association Records; Melvin Bork conversation with Shoptaugh, September 2007.

[6] Fenelon interview with Shoptaugh; Walker, *Combat Officer*, pp. 31-36. Several men commented in interviews that the enemy was detectable by an odor, exactly what they couldn't say. Some thought it was the gun oil the Japanese used on their weapons; others said it was the enemy rations, that the odor was "fishy smelling."

[7] S2 Journal entries for November 4-7; Baglien, original diary entries for November 4-6, 1942. A copy of Moore's order to Timboe, to "push NORTH to the sea tonight," was kept by Timboe and is in his daughter's possession.

[8] Rudolph Edwardson, interview with Shoptaugh, October 16, 2007.

[9] Walker, p. 36; Philip Engstrom's postwar account is written in his service book, following his diary. Walter Rivinius account in interview with Shoptaugh, November 2, 2007.

[10] James Fenelon's comment regarding Moore is from an interview with Shoptaugh, May 18, 2008. When Moore said "that was my mistake," he may have meant that no scouts had been sent ahead of the 2nd Battalion to warn the 3rd Battalion of its approach. A copy of Moore's order to Timboe, to "push NORTH to the sea tonight," was kept by Timboe and is in his daughter's possession.

[11] Colonel Shōji and his role in the battle is discussed in Frank, pp. 347, 353-54.

[12] In addition to Frank, pp. 422-23, see Miller, pp. 195ff., and Vandegrift, p. 195. There are no pages in the S2 Journal for the Koli Point actions from November 8-12. The official "Report of [the] Battle for Koli Point" (164th Infantry Association Papers), June 28, 1943, implies that "communication difficulties, as the wires were cut by the enemy repeatedly," was one cause for the failure to trap all the Japanese.

[13] Neal Emery interview, Barnes County Historical Museum. Joe Miller's story, written by John Stannard in 1960, is detailed in "Eightball Soldier," *164th Infantry News,* December 2000.

[14] Jim Campbell information in Cronin, *Under the Southern Cross,* p. 61

[15] John Paulson interview with Shoptaugh, October 20, 2007 and two subsequent telephone conversations in 2008 and 2009.

[16] In 1942, the U.S. Army had not yet mastered the tactical flexibility that the Germans called *Auftragstaktik* (literally mission command), in which a unit (squad or platoon for example) is given a clearly defined mission goal but then is allowed to develop its own tactics for acheiveing that goal. At the time of Koli Point, many company and even batalion level officers still gave detailed orders to small units, while others gave only the vaguest orders (see Moore's comments about officers in chapter 9). Effective ommunications between units was also still in development, as can be seen in Walker's remarks and in the fact that, on the night of 4-5 November, no men were sent ahead of the 2nd battalion column to warn the 3rd battalion of the 2nd battalion's approach.

[17] Walker, *Combat Officer,* p. 35. Walker's remarks about the death of Lieutenant Clark by an American mortar round suggests that 'friendly fire' was a danger that could have slowed the advance. John Stannard, "Eightball Soldier," *164th Infantry News,* December 2000. Stannard himself was awarded a Silver Star for his actions at the Henderson Field battle, and was wounded and evacuated in January 1943. He subsequently won an appointment to West Point, graduated in 1946 and remained in the Army until he retired as a Brigadier General.

[18] Miller, pp. 199-200. Timboe's relief is not mentioned in the regiment's "Battle for Koli Point" report, nor did Arthur Timboe mention it in a paper he wrote, "The 164th Infantry on Guadalacanal" (copy given to Shoptaugh by his daughter). Timboe also kept a notebook diary he wrote on Fiji, now in the possession of his daughter, which does not discuss the controversy. However, Maj. Gen. Bruce Jacobs, "Arthur C. Timboe of North Dakota: A Profile in Courage," *National Guard,* December 1996, relies on information that Jacobs received from Timboe.

[19] Vandegrift, p. 195-96. If Vandegrift had any views on the relief of Timboe, he obviously decided it was Army business, not his. Miller, p. 200, give the casualty count and estimates that "probably less than half" of those who escaped at Koli Point survived to reach the main Japanese base at Kokumbona.

[20] Patch may have anticipated trouble over Timboe's dismissal, and indeed the Governor of North Dakota and several other state politicos did complain to the Army. Moore himself was awarded a Silver Star by Patch for being at the front during the January battles for Mount Austin and participating in the evacuation of wounded men while under fire. Patch also supported Moore's promotion to Brigadier General after Guadalcanal. When Timboe returned to the states in 1943, he remained in the Army with various assignments in the states, Germany, and Korea. He retired in the late 1950s. See Maj. Gen. Bruce Jacobs, "Arthur C. Timboe of North Dakota: A Profile in Courage," *National Guard*, December 1996.

[21] Miller, pp. 200-202.

[22] Miller, pp. 177-80; Baglien diary, entry for November 12, 1942.

[23] Remarks of Richard Stevens from telephone interview with Shoptaugh. For detailed accounts of the naval and air battles, see Frank, pp. 428ff. and Eric Hammel, *Guadalcanal: Decision at Sea: The Naval Battle of Guadalcanal November 13-15, 1942* (New York: Crown Publishers, 1988).

[24] John Hagen remark in his dictated tapes; Clement 'Murphy' Fox to Marie and Leonard Schultz, November 15, 1942, letter courtesy of Rita Fox, Murphy's daughter.

[25] Miller, *The First Offensive;* pp. 204-206; Cronin, *Under the Southern Cross,* pp. 65-68; S1 Journal entry for November 20, 1942,

[26] Photographs in Miller show the hills and some of the bunkers.

[27] S1 and S2 Journal entries for November 21, 1942; Baglien diary for November 21, 1942

[28] S2 Journal entry for November 22, 1942; Miller, p. 208. Richard Stevens was commissioned after the war and rose to the rank of Colonel. He regarded the Matanikau offensive as a "very poorly conceived" plan created by Sebree and his staff. He also believed that the "forty yard" gain recorded for November 21 was "very generous; I doubt we got that far."

[29] Baglien diary, entry for November 25, 1942; Frank, *Guadalcanal*, pp. 493-97, summarizes the November battles for the hills.

[30] Richard Stevens, "Life on the Limbo Line," *164th Infantry News,* March 2009, and Stevens telephone interview by Shoptaugh.

[31] Duis story original in *164th Infantry News,* May 1999 (originally in the *Crookston Daily Times*, May 1943). Duis became a lawyer after the war and practiced in Fargo, ND, but was murdered by a prowler who broke into his home.

[32] Lester Aldrich information from interview of Aldrich by Merry Helm, 2007; Woody Keeble information from Helm, "Master Sergeant Keeble, Hero", Dakota Datebook Broadcast on Prairie Public Radio, May 16, 2005 (available at http://www.prairiepublic.org/programs/datebook/bydate /05/0505/051605.jsp). Helm has completed a biography of Keeble, who in 2008 was awarded the Congressional Medal of Honor for acts in Korea, 1951.

[33] Captain Andrew Panettiere, killed in action on the Matanikau ridges, was honored by the Army when one of the hospitals on New Caldonia was named for him; see "Order of the Americal Division Commander, December 17, 1942," in "History of the 164th Infantry, October 8, 1942 – June 30, 1943," p. 37, in 164th Infantry Association Records.

[34] William Tucker interview with Shoptaugh, September 16, 2007; Baglien diary entry November 30, 1942.

[35] Engstrom narrative in his diary; Rivinius interview with Shoptaugh.

[36] Baglien diary, entry for November 30, 1942.

Chapter 8: "Malaria, Exhaustion, Malnutrition"
Guadalcanal: November-December 1942

"The Matanikau is a bitch, a real bitch," a marine officer told Harry Wiens when Company I came across the river on November 20. Assaulting the hills beyond the river was a nightmare. The Japanese were well entrenched, the terrain was steep, the ground invariably rain-soaked and muddy. Men had to literally scramble and claw their way up wet slopes while under fire, only to lose their footing and slide down the way they came, and have to do it all over again. Snipers in well-disguised nests, let soldiers pass and then fired at them from behind. Enemy machine guns fired at any sign of a group of men moving through the tall grass. Reverse slopes were honey-combed with nests that could not be hit by long-range artillery.

Company I's Matanikau battle began with Colonel Hall calling the company commanders together to discuss the plan "to advance by fire and maneuver [moving] westerly approximately five miles to the vicinity of the Poha River." Attending the meeting with Captain Knott, Wiens was sitting and looking at the map while Hall pointed out the initial objectives, when he heard the "warning sizzle" of an incoming mortar round. Everyone quickly flattened out as the shell exploded. Well, almost everybody had hit the deck, for as he looked up while the dirt settled, Wiens saw Hall "sitting there, cross-legged, with the map before him; impassive, motionless, as if he were a statue." Surely, Wiens thought, with this man leading them, they would win this fight too.

But a week later, Wiens knew that even iron men could be hurt and that the marines had been right, the Matanikau was a killer (see Map 5). First, Hall was wounded in the face and neck by shell fragments. He was evacuated. A few hours after, Ralph Knott was wounded. He had had a close call just days before, standing on the ridge with Al Wiest, looking at the enemy positions. Wiest looked at Knott, who was about 6 and-a-half feet tall. Wiest thought he "made a lovely target." "We better get the hell off this ridge," he suggested. Knott turned his head to reply and just then a sniper's bullet zinged by creasing his neck slightly. "If Ralph hadn't turned his head at that very moment, he would have been drilled right through the neck."

Now, as the attacks on the hills bogged down, Knott's luck ran out. Harry Wiens had run back to the I Company CP to get a medic for his friend Christian Montgomery. Montgomery had been hit by a machine gun bullet that "laid open a deep furrow on the right side of his head." Knott saw Wiens at the medical tent and walked over to give him further orders. Suddenly Knott caught a bullet "in the left side of his neck . . . the force of the bullet spun him around, and the medic I was trying to corral attended to the Captain instead." Wiens carried both Knott and Montgomery back to the field hospital. Montgomery died, but Knott survived and was evacuated.

Wiens watched Company I melt away in nine days: "Captain Panettiere, our Battalion surgeon, and Lieutenant Whitney killed; Macey Paul, our 1st Sergeant, wounded, and Corporal Kohnke also killed . . . Lieutenants Sloulin and Grytness had been killed that first afternoon on the hill, and both Dibbert and Clewitt were dead." Wiens had spent a pleasant leave in the states with Clewitt, who told him "stirring tales of adventure and amorous conquest." As he was dying, Clewitt asked Wiens to write his girl friend and 'tell her I thought of her anyway.' "Even today when I think of him, my eyes still moisten," Wiens admitted. By November 29, the 3rd Battalion was so decimated that when Moore combined its companies together with the survivors of 1st Battalion's A Company, he totaled just 124 effective men, including the sick at heart Wiens.[1]

Company A and the rest of the 1st Battalion had been on the Matanikau front since October. Dennis Ferk remembered that in the first weeks, the main enemy was not the Japanese, but the tropical mosquito. "The mosquitoes there seemed larger that any I had seen and there were swarms of them." Like the other companies, the men of A were given Atabrine. "That turned our skin yellow and I stayed that way for a long time." Keeping dry was impossible. "Our foxholes were on our side of a big hill [Hill 80] and the holes were about four feet deep. We had shelter halves over the tops to keep some of the rain out. It rained all the time, and I remember bailing our hole out with my helmet." With so much rain, it was ironic that many men were taken to the hospital for dehydration. "It was so hot and humid; we lost salt all the time. We had salt tablets but our clothing was just white from the salt we leeched."

Men were lost steadily to sniper fire and shell bursts. "Rueben Heer, one of the Heer brothers, crawled up to the hilltop to take a look and we noticed he was just lying there," Ferk remembered. One guy went up and grabbed his leg, and he had a bullet hole right in the center of his head." Ferk lost several friends to artillery and snipers. By early December, his company was at about fifty percent effectiveness.

Company D got its turn in the Matanikau meat grinder. Assaulting the hills was PFC Vincent Clauson's "baptism of fire." When Clauson went forward on November 22, he and his friends passed a "pile of wrinkled clothes on the way up." Getting closer, he realized that the clothes were really the bodies of dead soldiers. He knew then "this was no picnic." Lugging his machine gun, he struggled to keep his footing up the slope and over the ridge. The Japanese found the range almost as soon as the men appeared on the ridge top, their machine gun bullets whizzing through the tall grass and forcing the men to crouch down. Clauson thought "it was a disaster to send us over the top of the hill. Jap mortars hit us and a piece of shrapnel hit me in the face. I could hear it coming and it hit me with a lot of velocity." His cheek was torn open and bled copiously as he tried to use his field jacket to stem the flow. "My buddy Arnie told me to roll over the hill and go down the other side on my backside. A couple of medics used bandages to stop the bleeding and then told me to walk to the aid station." Clauson got a Purple Heart and had face scars for the rest of his life, but he felt lucky all the same. "About six of our platoon were killed that day, a terrible waste."[2]

After the victory in defense of Henderson Field, the regiment's men were filled with confidence. Two letters written at the end of October illustrate the sense of triumphant optimism. Art Nix wrote to his family, October 31, 1942 -- "The food is okay, the cooks are really doing a fine job. . . Right now we are working as medics. It isn't too bad [remainder censored out]. In time we will get back together again as a band." John Hagen wrote to his father on November 3, 1942 -- "Right now, your season for ducks and pheasants is open. Ah, me, you have nothing on us. Some of the boys have their day's limit before sundown . . . We're doing our damnedest to clean up this situation as soon as we can."[3]

But such high spirits faded fast on the Matanikau front. "A lot of us had problems with malaria, exhaustion, malnutrition, you name it," Hagen said by mid-November. The death of Rilie Morgan, his chum from college, hit him hard. Morgan was killed during one of the assaults on Hills 82 and 83. "Rilie led another attack, calling out 'follow me' and heading up the hill," recalled Dick Stevens, who watched the forward movement unfold. "He got about fifteen yards ahead of his men and ran up to the hill top, but there was a sniper right there in the grass who shot him." Hagen bitterly blamed "throwing tired men into a headlong attack."

The band men and medics who went out to help the wounded were overtaxed, and Hagen volunteered with several other men from the 3rd Battalion headquarters to help carry wounded back to the rear. It was dangerous work, well described by Hagen's boyhood friend Bill Boyd: "The only way to get them was over the ridges and the Japs were shooting at you from both sides. You'd get up on a ridge and they would zero in on you, so you would go to the other side of the ridge. They'd shoot at you from that ridge so you'd go back to the other side." Four men would haul a man back with a blanket, slipping on the trails, dodging fire, praying and cursing in equal measure. Hagen and his mates, Ed Goff, and Bill and Ed Burns, carried several men for which they were later awarded Silver Stars.[4]

John Kasberger was also engaged in the Matanikau assaults, firing his machine gun in support of the infantry advances. "You were lucky if you could see a bunker, they were so well camouflaged. I fired on patches of vegetation in response to the sound of their own guns, or at a bit of smoke. It was frustrating. At night on that front it would rain and you got good and wet and you'd feel cold. One container of the ration packages was what they called 'waterproofed' with wax. You could burn that with almost no smoke, but it didn't really warm you. Sweating in the day, cold at night, you actually lived like an animal up there. Men were getting sick by then with malaria and jaundice. That took out more men than battle."

Art Nix was busy treating the sick and wounded. The regiment's bandsmen were regularly under fire, treating wounded and hauling ammunition. Nix, like Hagen and Kasberger and all the others, was very tired, but he never complained. In fact, he thought his duties were preferable to the buglers who had funeral duties. "Couldn't hardly blow taps as [they] would just choke up (no spit)," commented one man at a service. Altogether, the band men on the island had a rough go, and many castigated Gerald Wright, who had told them back in the states that they would never be in the combat zone. When Wright, sick in body and at heart, was sent back to New Caledonia in January, Art Nix wrote to his parents that "Wright isn't with us anymore, his nerves were in bad shape."[5]

Without the bandsmen, the American medical services on Guadalcanal would have been overwhelmed. Naval corpsmen had accompanied the First Marine Division when it landed back in August. These dedicated men did a fine job but there were not enough of them as the campaign developed. In early November, a much needed addition arrived in the form of the 101st "Provisional" Medical Regiment (later re-designated the 121st Medical Battalion). Part of the original Task Force 6814, this unit had been slated to sail to the island in October with the 164th, but limited shipping had delayed it for nearly a month. When they finally arrived, the medics and doctors went right to work establishing a second field hospital that included underground wards. Within hours, the unit historian noted, "the casualties soon came pouring in . . . Officers and enlisted men worked all day and most of the night, and did a splendid job under adverse conditions," conditions which included enemy shell bursts among the tents and the occasional infiltrating sniper. Two pressing problems that the hospital personnel tackled were the lack of whole blood and the shortage of good water. The 57th Engineers helped dig wells for the water problem. To alleviate the blood crisis, orderlies, medics and service personnel donated blood.[6]

Tech Sergeant William Dunphy was the medical regiment's supply disburser. He did anything he could think of to outfit the hospital. "We set up near what they called 'Tojo's ice house,' where I remember one cook who used it to make ice cream for the command level guys. We were well supplied for medical needs and had items that places in the states were short on, some surgical items and things. But we needed more lumber, sheets, generators and such. Well, we had quite a bit of 100 proof medicinal alcohol and I decided we could trade some of it to get that stuff. My commanding officer warned me, 'you could end up in Leavenworth for this,' but he turned a blind eye to it." Dunphy got what he needed from the Seabees and others. Some of the hospital's first patients were members of its own staff. Dunphy had counseled his staff to keep down during bombings but a few would "go out to look around and get hurt. Two guys who decided to watch the fireworks were killed when they came on a bomb that had landed in a tree and messed with the trip wire."[7]

Within a week of completion, the hospital took in 499 patients, many of them men of the 164th. Additional surgeons flew in from Noumea as casualties grew with the Matanikau offensive, and sometimes they performed near miracles. Paul Fischer was in the line with the rest of Company K, when his friend Harry Bennick was struck by a bullet. "We were trying to get close [to a machine gun nest] to throw grenades, Harry stood up to throw and he was hit in the head." Fischer carried Bennick back to the hospital, certain he would die. But the surgeons removed the bullet, stabilized Bennick and shipped him home. "He got back to the states and he got a plate put in his head and he was okay."[8]

PFC John Anderson was a medic with the medical regiment. Raised in Massachusetts, Anderson had grown up with a love of the ocean and was thrilled to traverse the Pacific with Task 6814. He felt it was exciting to set out for Guadalcanal, but soon banked his enthusiasm when he debarked from the troop ship in November; his boots tied around his neck to keep them dry, and passed "battle-fatigued and exhausted marines" waiting on the beach to be sent back to Noumea. "You are welcome to this Goddam island!" one of them called out. During the first few days, Anderson was at the hospital treating sailors who had been wounded and burned in the naval battles offshore. Then he went into the hills in front of Kokumbona to bring out the wounded men. Medics at that time had precious little in their field kits to help the men, mostly pressure bandages, sulfa powder, some ointments, and morphine ampoules. Anderson and his mates would treat the men as best they could and then haul them back to a field station. "John was a gentle man and did not [after the war] dwell on the horrors he experienced," his wife later noted. "He simply said that after a while you became immune and stayed focused on the job you had to do – help your wounded buddy stay alive with first aid and rush him to the field hospital. He always felt that it was hardest to get the men out of the hills."

Philip Rackliffe, another medic attached to Company B of the 182nd Regiment, agreed that trying to treat wounded men in those hills was "something like being in hell, trying to stop the bleeding of a shocked, hurt kid, while being afraid of getting hit yourself." Like Anderson, Rackcliffe wore no red cross insignia – "that'd just draw some sniper" – and carried a rifle to protect himself. He never used it but felt better to have it.[9]

By early December, over 1400 men had been treated at the army hospital, mostly casualties from the Matanikau hills and from malaria. Malaria, in fact, felled more men than the Japanese; the Army had yet to learn that officers had to strictly enforce the use of Atabrine to suppress malaria. The bandsmen, for example, were given little instruction on malaria prevention. Instead they were formed into four teams of litter bearers, attached to the three battalion aid stations and the regimental station. Douglas Campbell, a Staff Sergeant from Lisbon, related how a Service Company truck took his team to their location. "The marine detail was just finishing up their job [to remove brush] when we arrived. What do you suppose they did? They set fire to the brush piles and departed. [Sergeant Franklin] Schoeffler [Campbell's superior] was furious and yelled at them calling them a bunch of dumb bunnies (Sheff wasn't much for cussing). He told the marines the smoke would draw artillery," which soon followed. Schoeffler pulled Campbell into the already moving truck as the shells came down.

Campbell treated his first wounded man at Koli Point – "some action took place and some wounded men came to us and we treated their wounds. The man I treated had been shot high in the back of the shoulder blade area. I packed the wound with sulfa powder and directed him to where he could find a medical officer. He could walk ok and didn't need a litter." A few days later, he was leading a litter team along a telephone wire, trying to reach wounded men at a recent combat site, and got lost when the wire turned down a trail he did not recognize. Retracing his steps, he found the right trail. Only much later did he learn that the Japanese had laid a piece of captured wire down in order to lure the Americans into an ambush.[10]

The Matanikau front left the regiment bloodied and battle hardened. The situation there was stalemated after the failed assault on November 23. The

1st Marine Division assessment reluctantly admitted that "further advance would not be possible without accepting casualties in numbers to preclude [success]." Having taken enough of the hills to protect his perimeter Vandegrift decided "to hold what I had." His marines were worn out and due to leave the island as soon as General Patch took over control of the campaign. Patch arrived on the island a few days later and set up his headquarters in a wooden shack built by Japanese prisoners. He then went up to the Matanikau, visiting the company posts and asking questions. By the end of the month, he was convinced that he would need three full divisions to finish the job.[11]

Meanwhile, the two sides held their ground. "The American and Japanese forces now faced one another at close ranges," the U.S. Army history of the campaign notes, "the Americans on high ground, the Japanese on reverse slopes and in ravines." Francis Sommers, who had fought at Coffin Corner, recalled that they "waited for one month on that line." Sommers and his mates missed John Stannard who had been wounded and evacuated. On a patrol Stannard had spotted an enemy machine gun and then was hit in the face. The bullet "hit in one cheek and the bullet came out the other cheek." As one of Sommers' buddies put it, "the guy was always talking, no wonder it didn't hit his teeth." The effervescent Stannard wrote his mother that he had "a couple of small scars, but then I never did rival Robert Taylor for looks anyway."[12]

The two Americal regiments consolidated their positions, rotating companies on the ridge line at regular intervals, using aggressive patrols to safeguard against Japanese attacks, or locate more enemy positions. A form of static warfare prevailed, dominated by patrols, artillery exchanges, and quiet stretches punctuated by sudden, brutal fights. Casualty rates fell but the danger was still great. When a man in Neal Emery's platoon (G Company) was wounded in the back by a sniper, Emery managed to get a landing craft to the beach to take him back for help. "We were heading back and the motor quit, we'd run out of gas." The craft began drifting west toward the Japanese lines. "Pretty soon one of our planes came over and I thought he would fire on us, but he flew off. Then another craft came out a few hours later and towed us back. "That boat's pilot told us we were lucky because we were getting close to the Jap headquarters area." Emery took another risk a few days later when he found a shell buried in the sand. "I was afraid it might go off if other shells came down, so I dug it up, it was about three inches around, and carried it over to the water and dumped it. Thinking about it now [in 2005], that was stupid."

One of the eerier fights during this time was a day-long duel that Ralph Oehlke fought against a persistent enemy machine gun that was hidden on one of the beached Japanese ships off Point Cruz. This gun opened fire on any American who came into sight. So the Anti-tank Company brought forward some of its 37mm guns to knock out the enemy nest. At the Henderson Field fight, Oehlke had clearly seen the faces of the enemy as they charged the barbed wire. But this fight was a long distance affair, Oehlke and his men trying to spot the enemy gun hidden somewhere in the hull of the wreck. It was a bit like trying to pick off a duck with a single shot rifle. "We fired dozens of shots. Our gun wasn't recoiling properly, so after each shot we had to shove the barrel back into position." Sometimes the Japanese would stop firing, leading the gunners to believe they had knocked it out. "I guess we finally got it, but it took a long time."[13]

Night patrols beyond the Limbo Line were tense affairs. There were several types of patrolling, including combat patrols that involved as many as two hundred men. Most patrolling at this stage was for the purpose of getting prisoners, capturing documents or spotting enemy emplacements that could be attacked by artillery. This type of work invariably ended up behind enemy lines. Harry Wiens recalled many "forward scouts [who] were killed at this time," including one friend who, while his life ebbed away, plaintively mumbled "I never wanted to be a scout anyway." Dennis Ferk recalled a patrol into the hills south of the active combat zone. "There were about twenty of us, and we had a native guide. We found a small path and followed it. That night we slept about five, ten yards off the path, if you slept at all. We had our legs facing the path, so if someone came along we could just pick up our rifles and fire." The next day, the path took the patrol up a small knoll where they found four or five dead Japanese. "We turned their bodies over with rope in case they were booby-trapped. One had a diary with some sketches in it. I think that helped us locate several of their guns."[14]

Doug Burtell went on several patrols to get maps back to the S2 (Intelligence) section. "Our S2 officer would attach us to combat patrols and one of us would go along. You always travelled light, made sure that everything was secure so that no noise would give away our position. The guy in the lead would never walk down the center of the path. That was a quick way to get shot. He always went off the edge of the path so he could dive into the brush if the Japs were waiting and fired. We got papers from patrols and would translate them. I was struck by how flowery the diary entries could be." No one could be certain where a given trail would lead. But when Bill Welander went down one trail searching for enemy guns, he and the other man came across Japanese soldiers bathing in a small stream. The others asked Welander if he could throw a hand grenade far enough to hit the bathers. Having played some minor league baseball, he "thought I might just be able to make it." So he reared back and tossed the pineapple as far as he could. "We took off, but I think it got pretty close because we heard a lot of yelling after it went off." Once safely out of range, they laughed their heads off.[15]

A more serious patrol was led by Lieutenant Fred Flo and his reconnaissance platoon. This group of fourteen men, accompanied by a native guide and two marines, sailed a small boat to the west coast of the island and made contact with Father Emery de Klerk, a Belgian missionary who ran a small mission at Beaumont Bay. From de Klerk they learned that there were no Japanese anywhere close by. Flo verified the information by going on to Tiaro Bay, then returned to the American lines convinced that an American force landing on the west coast of the island could flank the entrenched Japanese positions. But by the time he got back, Vandegrift was handing command over to General Patch.[16]

Patch took command on December 9, 1942. Vandegrift and the exhausted First Marine Division left Guadalcanal. Patch was for the moment both the commander of the Americal Division and the commander of the XIV Corps making up all the units on the island.

As soon as more infantry arrived he would hand the Americal over to Sebree. His orders from Halsey were simple and direct: "eliminate all Japanese forces on Guadalcanal." Placing his Corps HQ in a large wooden shack by the Matanikau River, he went to work. Among his first actions as a corps commander was adding two disaffected men of the 164th to his staff: Arthur Timboe, former head of the 2nd Battalion, and Jim Beaton, a headquarters sergeant who had had serious run-ins with Bryant Moore.

Beaton had served in the S2 section since arriving on the island, but early on had decided that Moore was a "dirty guy" who "lied to us all the time." Moore called the men "second rate soldiers." Beaton also said that "Moore wanted the Seabees to build him a separate latrine because he was too refined to use ours. So the Seabees did it, but decorated it with fancy paint to show their contempt for him." In December Beaton became friends with a lieutenant colonel who served in Patch's XIV Corps HQ. "We were on a long patrol together and I got to know him. He suggested that I could do some good things in the Corps intelligence unit." The transfer went through and Beaton did good enough work to be offered a spot in the Officer Training School (OTS) program. Art Timboe was pleased to be in the XIV Corps operation. "I have an extremely interesting job and like it very well," Timboe wrote to his wife. "For one thing, we eat much better and the Marines and the other people are a fine bunch of people. The old gang is still here in the same place and I get to see them from time to time."[17]

Two battalions of the 132nd Regiment, the last part of the Americal, arrived on the island in December, giving Patch the combined forces of his division and the Second Marine Division. He decided to try and flank the Japanese entrenchments by swinging south to take Mount Austen and force the Japanese to retreat. But when the 132nd tried to seize the high ground, they were taken under fire by enemy guns from hills that the enemy had prepared for just such a move. They took over three hundred casualties. Stymied, Patch now decided to wait until the 25th Division arrived before commencing a new assault. He and his planners had rejected the idea of a landing on the west coast as "too risky." Until Patch could assemble the forces he wanted for his next attack, the Japanese were kept under artillery, mortar, and aircraft bombardment. Over time, the entrenchments that blocked an advance were, in the words of the intelligence officers, "progressively weakened" and could not easily be rebuilt.[18]

The daily routine on Guadalcanal was a mixture of long hours of tedium punctuated by brief moments of terror. Warren Griffin, a G Company combatant, found the mud as wearying as combat. "You couldn't see with the vines hanging down, you were knee deep in mud most of the time." Carl Tait, a private in Company G, took cover under a Bren carrier during one shelling, but the shell impacts caused the vehicle to sink further into the mud. "I needed help to get out." Virtually every man at Guadalcanal mentions mud as one of the things they most hated. Art Timboe wrote that "the earth melts into slime beneath the lash of tropic rains." "It clung to you like another skin," said one soldier. "You were never clean," said another. Men tried to clean themselves and their clothes in the rivers, taking care to post guards against Japanese snipers, or crocodiles. On one occasion Corporal Walt Ensminger, Jess Bandow, and Ted Kurtz, all part of the 3rd Battalion Headquarters, went down to the Lunga River "to scrub off some tropical sweat." When they got to the river, they saw "two old grey-haired gents" already bathing in the water. They ignored the men as they stripped down and jumped into the water. "Jess broke out a bar of soap and we lathered up," Ensminger remembered. Then one "old guy yelled over and asked if we'd share the soap. We were finished and tossed it to them." A few minutes later the older men climbed out of the water, dressed and pulled out of the trees "in a jeep with a metal two-star flag." It was then that Ensminger recognized General Vandegrift, driving off with "a bird colonel." He immediately told Jess Bandow who it was. Bandow stood there for a second and then said, "You know when he asked for the soap I just about told him to kiss my Irish ass."[19]

A visiting journalist recorded watching GIs wash their clothes in the Matanikau while Japanese guns dropped shells into the water about two hundred yards upriver. "I figure three more batteries [of guns] are needed to turn this river into a real washing machine," a tired soldier commented.[20]

When the men weren't trying to kill and stay alive, time crept with interminable slowness. Lieutenant Ray Ellerman, a band member who was transferred to Company L after going through an OCS program on New Caledonia, noted how "during the quiet times, we were always looking for something to read or do. We'd talk a lot, telling the same stories about home or what we'd do after the war." Jim Fenelon remembered that comic books had a great popularity among some of the men, but others read novels, histories, newspapers, anything they could get their hands on. A couple of units printed small news sheets; very few copies of these exist today. Baseball games were popular. A lot of men played cards, especially in the air raid shelters. Men who had made money from gambling or selling Japanese souvenirs to sailors would sometimes get money orders to send home. Harry Wiens remembered once taking "6000 dollars in twenties" to the camp post office to buy money orders for his buddies.[21]

There were no USO shows on Guadalcanal that year. Most such shows were still in the states with a few performers doing their routines for troops in North Africa. But one movie star came to the island in December. Joe E. Brown decided after Pearl Harbor that he would go out and visit American troops and paid most of his expenses while coaxing some officers to scare up transportation. After his son, an Air Corps flier, was killed in an accident, Brown made "the boys overseas" his cause for the duration. He went to Guadalcanal and did about a dozen shows there in three days, one of them seen by Dick Stevens, in a hospital tent where Stevens was fighting malaria with "the Cure of the week – liquid quinine." Even through the ringing ears that came with quinine treatment, Stevens marveled at Brown's routines, jokes and impressions. "For a half-hour or so of magical moments, we forgot about the weather, the hospital, our ailments. We were transported instead back to a world we thought we'd left behind – we few and Joe E. Brown."

Brown himself kept indelible memories of Guadalcanal. Visiting sick beds, he talked with many wounded men, including "a pretty sick boy" who had lost both legs. "I feel fine," the man said. "He couldn't lift his head and his legs were missing, but he felt fine." Brown used his gags to coax a smile out of the boy and at the end, the fellow said, "thanks for coming." "Aw, hell," Brown replied, "you'd do the same for me, wouldn't you?"[22]

One advantage that National Guard units had over other army units was the long standing familiarity and friendships among the men. Those who had been wounded wanted to rejoin their friends. While he was recovering from his wounds on Efate, Bill Tucker tried to bribe the doctor with a box of cigars to let him rejoin the 164th, but the doctor refused to sign the papers. Harold McClintock of L Company was evacuated with a shell splinter that had caught him in the face and caused a serious infection. Treated at a hospital on Fiji, he was released and assigned to a truck unit stationed in Suva. McClintock wanted to return to L Company, so when he heard that his friends Stan Anda and Eddie Burns were about to board a ship for Guadalcanal after recovering from their own wounds, he took his truck out to the Suva docks and went looking for them. Finding them he asked if they were going back to the outfit. Burns said, yes, they were leaving within a few hours. "Good," McClintock replied, I'll go, too," abandoning his truck at the harbor and boarding the ship. When the ship's captain received a radio message that McClintock was to be returned for desertion, Burns and Anda talked the captain into ignoring it.[23]

The men fought the loneliness that war engendered by sharing news of home. As Bernie Wagner of G Company put it, "We were all stuck on a primitive island, with nothing, no civilization, no shops, no family. What we had was one another, and the fact that many of us came from the same towns." By December, men had better mail service. And even after the censors finished going over their letters, there was news to share. Ray Ellerman remembered that "everyone who wrote home gave news to share with other families."[24]

Some of these letters, published in the local press, spoke movingly of the loneliness of war. Samuel Baglien wrote a letter to his local paper, the *Hillsboro Banner*, published on January 15, 1943: "We really have a group of men from our part of the country that we can be justly proud of. War at its best cannot be won without casualties. You at home undoubtedly realized the latter statement by now. To you folks at home who have lost loved ones, all I can say is keep faith for those you have lost as they have already kept their faith. Today is the day before Christmas and thanks to our friend, the quartermaster, we'll be able to enjoy turkey with all the trimmings. The boys have received a lot of packages from home and believe me, they are all appreciated."

Christmas Day was, in fact, rather quiet on the island with some patrols but almost no casualties. "Sure had a nice dinner," Baglien wrote in his diary. "Turkey, mashed potatoes, cranberries, nuts, candy – all the trimmings." Russ Opat and his fellow cooks had outdone themselves, all the more so for men who had been forced to live on cold rations for much of the previous month. But as Les Wichmann noted, they still had to clean their mess kits "in the usual 3 GI garbage cans of hot soapy water." After the meal, Art Nix played a record that his parents sent him of a concert given by the Enderlin Kiwanis, over and over. "Just can't tell you how I felt when I heard it," he wrote home. Little things meant a lot.[25]

Mail was crucial for morale. When it came it was a treasure. Lloyd Fox was a native North Dakotan, from Schaefer, who joined the Navy in 1938. He was serving on the destroyer *Cushing* when it was sunk in the great naval battle on November 13. He and other survivors of the crew were picked up and taken to Lunga Point, and he spent over two weeks on Guadalcanal. After helping to unload some of the ships that brought supplies in, Fox was assigned to help out in the base post office. "I was helping going through the mail going back to the states, and I saw a letter addressed to Mrs. Joel Grotte. Joel was one of my old beer drinking buddies in Watford City! So I left a note for him and later that evening , here he comes. I didn't know they were there until then. We had a good visit for about an hour, talking about home." Grotte was himself recovering from wounds he received in the Matanikau fighting, some shrapnel in his back and a small piece that slightly injured one of his eyes. Hubert Flannery, one of the senior medical officers, smeared his eye with antiseptic and covered it with an eye patch. Grotte stayed with his unit in the combat zone until they were all sent back into reserve.

One of the most vivid yearnings for home was written in a letter by nine alumni of Rugby High School, including Bill Love, Bob Davidson, Wes Spillum, and Clarence Bednarz, all serving in Company D. After relating some of what they faced on Guadalcanal -- thirst, jungle rot, fever, and "slaughter" -- they concluded with the observation that having seen enough of the exotic South Seas, they wanted to go home and "settle down for good

. . . In all our travels, we've seen many strange lands and peoples and we wouldn't trade one old beaten up, trod-on snow bank in ND for all these tropical splendors." Poets couldn't have crafted a better image.[26]

Combat led young men to think about everything from friendship to mortality and faith. Combat soldiers harbored some resentment toward the 'rear-echelon.' Alvin Tollefsrud kept an indelible memory of his one visit over to the Seabee camp. "They lived like kings! They had running water and showers, good food, you just couldn't believe it compared to what we had." He also resented the "HQ boys," and regarded some of Moore's staff as nothing more than "dog robbers." Many men took it for granted that Moore's headquarters lived in luxury. That was hardly the case. There were no real luxuries on Guadalcanal. Officers did have greater access to alcohol and several of them used connections to get cigars and other comforts flown in on mail and payroll flights. But all men were equally exposed to illness and fever. Those closer to Henderson were often more exposed to air bombings. Artillery shells were indifferent as to who was hit by shrapnel. It was how a man behaved under fire that mattered most.[27]

As a pre-war guard unit, most of the men had known their officers personally as well as officially. But in war, what mattered was how the officers held up under pressure. Some of the company commanders failed to make the grade in the eyes of their men. Most of the enlisted men steered clear from Colonel Moore's argument with Baglien and others, but they expected a great deal from their platoon and squad leaders. John

Gossett, Ralph Knott, Chuck Walker, Miles Shelley, John Tuff, and Rudy Edwardson were some of those that the men respected. Gerald Wright had lost the respect of many of his men, both because his promises that they would not see combat seemed deceptive and because he himself took few chances.

The constant presence of death made men think. John Paulson certainly was not the only soldier who prayed during battle, nor was he alone in finding comfort from the Bible or church services while on Guadalcanal. "I was a weekly attendant at masses," Ed Mulligan wrote later, "and relied on my prayers for my safe return." The regimental chaplains held Sunday services, which were well attended, when combat permitted. The Christmas Day services were packed. A Signal Corps photograph of one of the Protestant services shows virtually the whole of Company G pressed in a tent to hear the portable pump organ.[28]

The men treated most of the chaplains politely, but the one man who held their highest esteem was Father Thomas Tracy. A native of St. Paul, Minnesota, Tracy was thirty-one when he went with the 164th to Guadalcanal. He had been with the regiment since 1940, and insisted on staying with it when the men went into Federal service. At Claiborne, the men had taken a real liking to him because he was what they called "a regular guy." Colonel Sarles was so impressed with Tracy's rapport with the men that he made him a captain before the unit went overseas. It was on the island that he earned the lasting respect of all of the men. "Father Tracy was every inch a soldier," Harry Tenborg told a reporter in early 1943. "He would say mass for the Catholic boys and then conduct services for the Protestants." He seemed fearless, making so many trips up to the front lines that the men started calling him "the foxhole padre." "Father Tracy repeatedly disregarded his own safety," Tenborg noted, in order to "offer guidance and encouragement to troops in foxholes and dugouts." At one point, Tracy ministered alone because the Protestant chaplains had been evacuated with malaria and fatigue.

Tracy could be relied on to help with just about anything, small or large. Harry Vadnie was sergeant in the HQ Company. He had joined the Guard in 1939, at seventeen, and had "exaggerated" his age to get in. But later, when his mother discovered that the record showed him as born in 1921, she was embarrassed. She and Harry's dad had married in January 1921. She wrote Father Tracy a quick plea: could he somehow fix the record? No problem. Tracy took care of it, and as Harry noted "I had been legitimized!" Jim Fenelon thought Tracy was unique because "he was good to everybody and had everybody's respect, even Bryant Moore liked him." Tracy was awarded the Bronze Star by Moore.

Tracy returned the men's respect. "The propaganda emanating from Tokyo, to the effect that we are totally cut off and isolated from the states and Australia," he wrote his sister, "are laughable." He told a reporter that Guadalcanal was won by "the quiet soldier who did his duty in civilian life and at home," rather than the "tough guys who talk out of the south side of their mouths." He told a Minnesota writer that the men who came home from the war would have "little complaint about life, little discontent, little inclination toward wanderlust." They would build a better nation instead.[29]

As 1942 drew to a close, a better world seemed as remote as flying to the moon. After a hundred days of combat, the 164th was worn down, the units reduced by men killed, wounded and sick. The men were drained beyond measure. Army news told them that over 250,000 Americans were in the southwest Pacific. When an Australian asked a debarking GI at Melbourne harbor "How many of you guys are coming," the soldier said "Boy, when we arrive [in force] you'll feel like [the] strangers." Indeed more than a million Americans would be in Australia by 1945. So why weren't more of these guys here, they asked on Guadalcanal. When would the Army send enough force to finish the fight for the "green hell?"[30]

(Endnotes)

[1] Harry Wiens, "My Little Corner of the War," pp. 120-31.

[2] Dennis Ferk, Interview with Shoptaugh, September 16, 2007; Vincent "Swede" Clauson, video interview with his nephew, Don Knudson, January 28, 2007.

[3] John Hagen to Olaf Hagen, November 3, 1942, Hagen Family Papers.

[4] John Hagen tapes; Richard Stevens interview with Shoptaugh; William Boyd interview with Shoptaugh, August 20, 1993.

[5] John Kasberger, telephone interview with Shoptaugh, August 30, 2007; memories of Art Nix and Robert Sanders, in *164th Infantry, IInd Island Command, 294th A.G.F. Band Reunion* booklet, October 1985.

[6] "Unit Journal, 121st Medical Battalion," June 30, 1943, RG 112, Box 255, National Archives.

[7] Shoptaugh telephone interview with William Dunphy, September 5, 2007.

[8] Paul Fischer, interview with James Fenelon, September 19, 1996.

[9] Frances Sinclair-Anderson to Shoptaugh, July 16, 2007, and telephone interview with Mrs. Anderson, September 12, 2007; Shoptaugh telephone interview with Philip Rackliffe, March 14, 2007.

[10] Douglas Campbell, undated narrative, in *164th Infantry, IInd Island Command, 294th A.G.F. Band Reunion* booklet, October 1983.

[11] "Report of the Battle of the Matanikau," June 28, 1943, 164th Infantry Association Papers; Miller, *The First Offensive,* pp. 208-209; Vandegrift, *Once a Marine,* p. 201; Cronin, *Under the Southern Cross,* pp. 69-71; Cooper and Smith, *Citizens as Soldiers,* pp. 289-90. William K. Wyant, *Sandy Patch: A Biography* (New York: Praeger, 1991). pp. 57-59.

[12] Francis Sommers, interview by James Fenelon, October 30, 1996, for story of John Stannard's being wounded; Stannard's letter to his mother reproduced in *164th Infantry News,* April 2000.

[13] Neal Emery interview with W. Michael Morrissey, November 14, 2005; Ralph Oehlke, interview for Barnes County Historical Museum, March 23, 2004.

[14] Wiens, "My Little Corner of the War," p. 123; Shoptaugh interview with Dennis Ferk.

[15] Burtell and Welander interviews with Shoptaugh.

[16] Fred Flo Patrol information from Cooper and Smith, *Citizens as Soldiers*, p. 288, and John L. Zimerman, *Marines in World War II: The Guadalcanal Campaign*, (Washington: U.S. Marine Corps, 1949) p. 152.

[17] Vandegrift, pp. 201-3; Cronin, p. 71; James Beaton, interview with Shoptaugh, June 4, 2007; Arthur Timboe to his wife, December 5, 1942, extract shared with Shoptaugh from Mary Fran Riggs, e-mail, November 2, 2008.

[18] Wyant, *Sandy Patch*, pp. 59-60; Cooper and Smith, p. 291. Good descriptions of the artillery's work at this time are in Bill McLaughlin's book, *The Americal Generation*. The decision to continue assaulting the hills was questioned at the time by some officers. Samuel Griffith, one of the Marine Raider commanders and author of a book on the Guadalcanal campaign, regarded the Matanikau-Mount Austen offensives as "a bloody fiasco." The proposal to flank the Japanese with a landing on the west coast was discussed but apparently rejected on the grounds that maintaining a separate base of supply would overtax the Navy. Halsey was wary of additional bridgeheads after the Aola Bay plan was scrapped and Patch's advisors still expected the enemy to attempt another landing backed by battleships and aircraft carriers.

[19] 164th veterans' comments, from Ed Murphy to Carl Tait, on conditions at Guadalcanal, are from interviews conducted (on videotape) at 50th Anniversary Observance Ceremonies at Camp Grafton, ND, October, 1992, now filed at ND National Guard Headquarters Offices, Bismarck. Ensminger story of Patch in "Generally Speaking," *164th Infantry News*, October 2000.

[20] Ira Wolfert, *Battle for the Solomons* (Boston: Houghton Mifflin Co. (1943), p. 138.

[21] Raymond Ellerman, telephone interview with Shoptaugh, September 19, 2007; Fenelon, telephone interview with Shoptaugh, April 26, 2007; Wiens, p.151. Wiens also noted that in one game he held a royal flush – and won "a mere seventy-five cents."

[22] Richard Stevens, "Joe E. Brown, 1942," *!64'Infantry News*, July 2008; Joe E. Brown, *Your Kids and Mine* (New York: Doubleday, 1994), pp. 101-2. According to his sister, Leatrice Cooper, Vince Clauson liked to tell people about meeting Brown on Guadalcanal and say "shake the hand that shook the hand of Joe. E. Brown."

[23] William Tucker interview with Shoptaugh, September 16, 2007; Thurston Nelson, "From Prairie to Palm Trees," (typed reminiscences of Nelson, a member of Company L).

[24] Bernard Wagner, telephone interview with Shoptaugh, August 10, 2007; Ellerman telephone interview with Shoptaugh.

[25] Christmas memories in *164th Infantry News*, December 2000 (Wichmann memory) and December 2001; Nix diary entry for December 25, 1942, *164th Infantry, IInd Island Command, 294th A.G.F. Band Reunion* booklet, Oct. 1985.

[26] Lloyd Fox and Joel Grotte, interviews by James Fenelon on September 21, 1996; *Pierce County Tribune* (Rugby) in February 1943. Neal Emery, a rifleman in Company G, in a short interview, February 15, 2001 (preserved at the Barnes County Historical Museum, Valley City) said that the war induced a yearning for home so strong that after he returned to the states he decided to avoid the greater world, a vow he pretty much kept by living on his family farm for almost 60 years.

[27] Tollefsrud interview with Shoptaugh.

[28] *164th Infantry News*, December 1998, for Signal Corps photograph of G Company on Christmas day; Edward Mulligan to Winifred Berntsen, May 7, 2001. Ms. Berntsen, a grand-niece of Thomas Tracy, kindly shared this letter and much other information with the author concerning Father Tracy.

[29] "Foxhole Padre Dies," Fargo *Forum,* October 21, 1960 (reprinting March 1943 remarks by Tenborg; Howard White interview; Shoptaugh interview with Fenelon; Harry Vadnie to Shoptaugh (e-mail) March 7, 2008; undated news clippings and Tracy letter to "Frank and Mat [his sister]," November 27, 1942, copies courtesy of Ms. Berntsen.

[30] E'D. Potts, *Yanks Down Under, 1941-1945* (New York: Oxford University Press, 1886), pp. 26-29.

Chapter 9: Bitter Sweet Victory
Guadalcanal: January-February 1943

"No food available . . . we can do nothing to withstand the enemy's offensive" read a late December cable sent to Tokyo from the 17th Army headquarters in Rabaul. Without greater support from their navy, the Japanese soldiers on the island were going to starve. But the navy declined to risk any more large ships, arguing that loss of more would make it impossible to defend the rest of the Pacific from the growing might of the Americans. American air superiority decided the campaign. One of the Japanese generals declared after the war that "the superiority of the American Army planes made the seas safe for American movement in any direction and at the same time immobilized the Japanese Army as if it were bound hand and foot." The fast P-38 Lightnings, and a new navy fighter, the Corsair, flying from Henderson, made short work of any Japanese pilots foolhardy enough to appear overhead in daylight.[1]

A captured letter, dated December 1, taken from the body of a Japanese soldier killed in front of the 164th lines, revealed the enemy's despair: "No sign of friendly planes or of our navy appears. The transports have not come yet either. I have not eaten properly since the 24th of November, many days I have had nothing to eat at all. From tonight on, indefinitely, again without expecting to return alive, I am going out resolutely to the front line. I must serve as long as I can move at all." Some of the Japanese officers on Guadalcanal asked for permission to attack in a last charge "and die an honorable death rather than die of hunger." The Chiefs of Staff in Tokyo denied the request and sent word that destroyers would evacuate the men from the island. The Americans failed to intercept this message.²

As the year began, the last units of the 25th Infantry Division were arriving, giving Patch about 50,000 men for a new attack, about twice the total enemy force. In addition to the 25th Division, Patch decided to employ the 132nd regiment and parts of the 2nd Marine Division, which with the 147th and 182nd regiments he formed into a "composite Army-Marine" division. The 164th was held in reserve. Uncertain how badly the enemy was weakened by hunger and disease, Patch's planners still expected the Japanese to try to reinforce their troops on Guadalcanal.³

When the 25th Division arrived on Guadalcanal, Art Nix and several other men from the town of Enderlin were surprised to be summoned back to headquarters. Leaving Company B's camp, Nix wondered what he had done to displeased someone back "at the top." But when he got there he was delighted to see an old acquaintance, George Christianson, also from Enderlin, now serving in the regimental HQ of the 27th Infantry, a regiment in the 25th Division. "George was a little older than us and he had become a major after being in ROTC at college. When he heard that there were guys from North Dakota on the island, he asked to see those of us from Enderlin. "He said he hadn't seen anyone from Enderlin in so long, he just wanted to talk to us. It's funny, we talked and we were all so happy that when we got ready to leave, we hugged him. His aide got upset and said 'don't hug the Major!' But George just said it was okay."⁴

Nix was busier now with illness than with wounded men. Malaria was seriously rampant in the 164th. Sam

Baglien was "diagnosed with moderate-severe [malaria]. I am taking six quinine and three Atabrines a day then three Atabrines a day for four days. Doc [Schatz] told me to stay at CP for a couple of days." He was also sick at heart because with Patch now in command of the XIV Corps, Sebree was made commander of the Americal Division. Moore was also promoted, an action that Baglien took notice of in his diary: "No air raids. Col. Moore made Asst. Division Commander. Lt. Col. [Paul] Daley, Asst. G2, our new [regimental] CO. This makes twice our own [i.e. National Guard] officers have been passed up. Guess you got to belong to the family."[5]

Moore's departure from the regiment eased the tensions in the regimental headquarters. Moore had removed a number of officers in a quick and forceful manner. Whatever niceties and politeness that may have permeated the Guard in peacetime, he spent no time on such things with lives at stake. In December, Moore was interviewed by a fact-finding staff officer from General Marshall's headquarters, seeking first hand information on how combat training should be changed. Moore told the officer that "[t]he biggest problem here is the leaders, and you have to find a way to weed out the weak ones. . . . The good leaders seem to get killed; the poor leaders get the men killed. I have had to get rid of about twenty-five officers because they just weren't leaders! I had to *make* the battalion commander weed out the poor leaders."[6]

Moore's actions were not unusual. In 1942 many officers were replaced for not making the grade. General MacArthur ordered the dismissal of dozens of officers during the New Guinea campaign, including a divisional commander who was a West Pointer. In 1943 the American army commander in Tunisia was replaced after his troops suffered a setback from the Germans. Failure or lack of progress generally led to the rolling of heads. It was the Army way. There were certainly officers in the 164th who did not handle things well on Guadalcanal. At least two company commanders were replaced for incompetence, one being transferred to duties in the Service Company because he failed to measure up. And as Sergeant Edwardson's remarks about the failure of Company F to push forward quickly enough at Koli Point suggest, some of the foot soldiers had their doubts about some of the officers. Doubts in war can lead to hesitation, mistakes and disaster.[7]

West Point graduates like Eisenhower, Bradley, Patch, many others, had waited decades for a shot at higher command. For every officer that George Marshall had marked in his mind for a major role in the war, dozens of other long-serving officers hungered for their opportunity to lead a regiment or a division. So it was no surprise that one West Point man helped another up the command ladder during the war. Bill Boyd watched this happen in the Americal and in the 164th, and said later that the Army "wanted to run their 'potential generals' in and give them some combat experience. They called us the stairway to the stars."[8]

Sam Baglien and several other Guard officers were convinced that only West Point officers (the "Army family" or the "ring knockers", or the WPPA – West Point Protective Association) were going to advance themselves and their friends during the war. There was truth to this; West Point looked after its own. But then, the Guard had looked after its own too, both in peacetime and in the early months of the war. But now Baglien predicted that none of the higher-ranking officers of the "old regiment" would get a higher command after Guadalcanal. This especially grated men who respected Bob Hall. Hall had the respect of Vandegrift and Puller and was "the ablest battalion commander" in the view of Al Wiest and many others.

How much the Guard officers' bitterness and sense of hurt contributed to their fate, is a question worthy of debate. Ultimately this dispute was a brief and minor dispute in the context of the desperate Pacific war. When the firing stopped in early February, Patch was grateful to Moore, and Moore no doubt suggested that a fresh team for the regiment and battalion commands was in order. Baglien and others were going to be sent home.[9]

Below level battalion, the 164th needed rebuilding. It had lost about half of its original company commanders and a great many lieutenants. Some of them, like Ralph Knott, were wounded or killed, others evacuated for illness or other reasons. In most cases, the replacement commanders were lieutenants who had made the grade as combat commanders. Knott's replacement was Ted Steckler ("Steck"), a 2nd Lieutenant from Company K. Harry Wiens was pleased with the choice, finding Steck to be a good and careful leader. Several sergeants who had done well were given field commissions as 2nd lieutenants, including Joe La Fournaise and Howard Van Tassel in Company I and Bernie Wagner in Company G. Wagner accepted the promotion but still felt that sergeants had the better situation. "You never thought of that as a promotion. When you're a sergeant, there's always another shave-tail coming in." Indeed some of the young "looies" who came to the island to fill a spot did not last long. Bill Welander picked up a lieutenant who had flown in to Henderson and drove him out to his battalion. Twice the man dove out of the jeep and took to the trees when air raid sirens went off. Welander told his CO that the man was "too jumpy" to be much good and he was right.[10]

By January, battle casualties and malaria had ravaged the 164th. Some of the men who were healthy enough went out on patrols or made up a perimeter guard in case of a Japanese raid. Many of the men were assigned to unloading supplies at the beach. The 3rd Battalion, with Colonel Hall back from hospital and again in charge, was assigned to coastal defense. Wiens was very happy to find Company I placed close to the beach. Situated south of the equator, Guadalcanal's hottest weather came in December and January, but the ocean breeze not only alleviated the heat it also discouraged the flies and mosquitoes. Wiens and his friends were finally able to clean up a bit, remove their beards, and enjoy the sun in t-shirts and skivvies. They also had tents for the first time. "We now had cots and mosquito nets. I really felt sorry for the guys [now] on the line and wondered how they endured without effective nets." In order to keep insects from crawling all over them at night, men put their cot legs into cans of water or kerosene.

Wiens was given a new job as the company communications sergeant, but he still went on several patrols, including one very long and important one in January. A Company G patrol had discovered the trail that the Japanese had cut in October for a disastrous attack south of Henderson Field. Regimental HQ wanted this trail thoroughly reconnoitered for fear the Japanese might use it "to mount a surprise flanking attack." Wiens volunteered a small patrol which would rely on speed; "on this venture we would use absolute stealth." Early on along the trail they found a "large artillery piece" in a gorge, and Wiens wondered about the "prodigious supreme effort" needed to get this gun so far inland only to abandon it short of the American lines. They found a "huge pile" of enemy equipment, about ten feet tall and "more than fifty in diameter." Along the trail sides were shoes, clothing, packs, and weapons. "This is where they tossed aside their equipment after their defeat at the Henderson Field battle," he decided. Further on there were "numerous skeletons" in "tattered [Japanese] uniforms." Seeing all this, Wiens and his comrades began to feel something of how the enemy had suffered. It "must have been agonizing, excruciatingly terrible, though at the time none of us felt any pangs of sympathy; they had only recently been attempting to dispose of us, and inflict a similar fate."

The next day, they followed the trail to higher ground and saw laid out before them the American perimeter and the airfields. Further on they encountered a large group of abandoned Japanese tents with more equipment and even "neatly stacked rifles." Each man hoisted a couple of Arisaka rifles apiece, which they could sell to sailors for $35 apiece. Wiens and the others then retraced their steps and made their report. The absence of "live Japs" on the trail was received as good news at XIV Corps, which had begun what it hoped would be the "decisive attack" on the enemy.[11]

On January 10, the 25th Division and the 132nd Infantry Regiment renewed the assault on the Japanese hill positions. The composite division of marines and soldiers meanwhile began advancing along the coast. For the next two weeks, soldiers and leathernecks clawed their way over the stubbornly held Japanese positions, destroying fortified positions with explosives, artillery, air strikes, and sheer determination. Island natives, organized into work gangs by British and Australian district officers, helped maintain the advance by carrying water, rations, and ammunition up into the hills. Other island workers enlarged the airfield, unloaded ships in the sound, and did other work that freed soldiers for combat duty. A special group of island scouts, organized by Martin Clemens and his native police contingent, acted as guides for patrols.[12]

The flamethrower was employed for the first time on the 15th when a team of marines used one against a pillbox by the beach. "Covered by automatic rifles, [two men armed with a flamethrower] crawled to within 25 yards of the position and fired the flamethrower at the bunker," noted an official account. "All resistance ceased, and the marines found 5 dead Japanese inside. Although 2 of the enemy had managed to get out, neither had escaped the effects of the flame." The same men destroyed two more pillboxes a few minutes later. Thanks in part to this new weapon, the advance continued. By January 23, the 25th Division had seized much of Mount Austen and the surrounding ridges. The enemy was being pushed back on the coast. But the cost was high, as some three hundred men paid for it with their lives. About twice as many had been wounded, while over 3000 Japanese fought to the bitter end and died in their gun pits. Meanwhile, Japanese destroyers were evacuating men by making high-speed night runs.[13]

Japanese prisoners were rare. Soon after he arrived on the island, Lieutenant Tony Hannel learned that Japanese warriors chose death over captivity. Leading a Company C patrol along the beach, Hannel was told by his scout that a man was in the brush about a hundred yards ahead. "Putting my field glasses on him, I could see that he was much too light to be a native." Then spotting a pilot's cap, he knew he had a downed enemy flier. His men approach cautiously. "The Jap aviator then ran to the seashore . . . We wanted to bring him back with us for interrogation, but after several attempts to take him alive, he bolted into the sea and started to swim away. We had no choice but to 'plug him' . . . I surely wish we could have captured him alive.[14]

Bill Welander took a prisoner in an unusual way. It was toward the end of the campaign. Welander was just returning from a reconnaissance patrol, and it was starting to get dark. As the men filed into the perimeter lines, one of guards asked Welander, 'hey, who's that behind you?' Turning around, Bill looked at the face of a small Japanese soldier, in a ragged uniform, obviously sick and exhausted. "He must have just joined us in the dark and walked in, probably hoping to find some food." The Japanese just stared at him, his face blank. After searching him for weapons and finding none, Bill took him over to the nearest first aid post. There, the doctor on duty became angry at Welander, saying something like 'why did you bring him here, just shoot him.' Bill stood there for few seconds, saying nothing, and then slid his rifle off his shoulder. "I just held it out to the doctor and said if he wanted the guy dead, then he could do it." A few more seconds went by, and then the doctor told a medic to look over the prisoner. Welander saw the Japanese again, in the small prison compound, and felt good that he had helped the man survive.[15]

Medic Gerald ('Sandy') Sanderson also had an encounter with Japanese prisoners. Due to a serious case of dysentery, Sanderson had remained in the hospital on New Caledonia and finally got to Guadalcanal in early December. Assigned to the 2nd Battalion aid station, Sanderson went everywhere armed. He and a friend visited the Japanese prisoner and talked to one of the MPs, who told them that the Japanese prisoners did little to help one another. They never helped a fallen man get back on his feet, he said, never went to the aid of a dying man. Sanderson's friend asked "I wonder if we'd be like that if we got in such a position?" Sanderson admitted that he did not know the answer.[16]

The regiment's routine had become a litany of work gangs, patrols, watch lines against Japanese infiltrators and beach watch. There were constant rumors: that another Japanese division was going to land on the island; that the Japanese navy was bringing all of its forces down for one last battle; that four carriers were leading the enemy flotilla; that all the remaining battleships were coming to shell the island again. No one would quite accept that the Japanese were being beaten now. Having been there for so long, the men seemed to feel the fight would just go on and on, that assuming continued stalemate was preferable to feeling first optimistic and then disappointed. There were not many casualties now, but some occurred. Bernie

Wagner came back from a patrol that had gotten into a firefight with a Japanese patrol. Wagner had lost his helmet during the fight and then left it to help carry a wounded man. The next patrol from G Company found his helmet and many men assumed he was dead for several days. Harry Wiens came back from another of his patrols, which was uneventful, but was saddened to learn that Clarence Bonderud, his friend since Claiborne days, had been killed by a Japanese night bomber. Popular, "slim and quick in his repartee and action," Bonderud was "a splendid, fun-loving fellow" who everyone had liked. Now his friends mourned his death.[17]

Casualties did not come solely from enemy action. There were a lot of different ways to get hurt on Guadalcanal. A few men drowned while crossing rivers or swimming on the beach. A man in Rudy Edwardson's platoon was badly hurt when American PT boats mistook Edwardson's patrol at Point Cruz for Japanese and opened fire on them. The man "was shot in both knees and had to be evacuated." A terrible accident killed several men in the 57th Engineer Combat Battalion, who had removed land mines from the perimeter and were trucking them back for disposal. The truck passed Alvin Tollefsrud and some of his buddies. A few seconds later it exploded. "We hit the ground the minute we heard the blast but you could feel the heat on your back . . . It was a big mess. I saw part of the truck frame later in one of the trees. All those engineers were killed, of course. There really wasn't much left of them to pick up."[18]

Illness felled most men, with malaria being the biggest problem. Gerry Sanderson spent most of his time treating sick men. "We had a lot of yellow jaundice, hepatitis. I think it spread from tainted yellow fever serum, shots that we all got." The doctors flew in "cans and cans" of hard candy for the most serious jaundice patients, "I guess to give the men more sugar." Dengue fever was rarer but could be more dangerous. Bill Hagen had a case of dengue fever: "With dengue you either live through it or you don't live through it." Hagen's case ran its course and his fever broke after several days spent "shaking and wringing wet" on a hospital cot. He remained weak for days. Almost everyone had some jungle rot, a fungal infection that if untreated could lead to blood poisoning. Sanderson treated "hundreds for jungle rot. It was easy to get. We were always wet. We had ponchos, but it was so humid, when you were marching you would get wringing wet from sweat." Sanderson treated the skin sores with lotions and urged the men to try and keep dry. "I still think fatigue and nervous psychosis contributed to the skin problems. The men would be worn out and scared and just keep scratching at their skin." Dennis Ferk and other men treated themselves with small bottles of Absorbine Junior. Howard White contracted blood poisoning from his sores. Fearing for his life, the doctors "put me on a C47 for Espiritu Santo [New Hebrides]." The airplane got lost in the clouds and it took some time for the navigator to regain his bearing. The prospect of crashing into the sea made White forget his infection for a while.[19]

Another problem, seldom discussed among the men, was combat fatigue. After weeks in the jungle, under fire, exhausted and dispirited, some men simply became unable to function. Sanderson remembered only one case that he dealt with as a medic. "This fellow had been a good soldier, outgoing and all, and then he was just very quiet, lethargic, didn't respond to anyone." Sanderson was later told that the man had shot a Japanese soldier who had only a rusted, useless weapon. "Maybe that's what did it." Bernie Wagner brought two men back from one of his patrols who had "cracked up from the stress." A number of officers took a dim view of combat fatigue and General Patch wanted "neuropsychiatric cases" court-martialed. But doctors usually listed such men as "victims of blast concussion and other organic injuries," treated the men with sedatives and let them rest. Most returned to duty.[20]

Malaria continued to spread. "About all of us got malaria," Sanderson noted, "although many didn't get really bad until we got to Fiji. We had a sergeant, called him an 'Atabrine noncom,' and he watched us take it. Some guys tried to dodge it because of the rumors it would make you sterile, when, of course, a real high fever from malaria *would* make you sterile." So far, the Army had put little into reducing the malaria threat by draining stagnant water pools. There was only one such team of sixteen men on Guadalcanal until the very last weeks of the campaign.[21]

The last stubborn stand of the Japanese defenders came at the Bonegi River, at the end of January. While a select unit of infantry held on there, destroyer-transports pulled nearly 5,000 men off the island and ran them north to Bougainville. The sailors were shocked at the emaciated appearance of the soldiers they carried. U.S. intelligence meanwhile interpreted the Japanese naval activity as a sign the enemy was trying to bring in reinforcements. Still worried by this prospect, Patch had dispatched C Company of the 164th to Savo Island to see if any Japanese forces were hiding there. But after a week of searching, the soldiers found nothing. After the defenses on the Bonegi were broken, the Japanese began falling back toward Cape Esperance, making stands just long enough to hold up the Americans while the evacuations to Bougainville continued (see Map 6).[22]

With Mount Austen and the hills around it neutralized, and the enemy falling back along the coast, patrols pushed northwest to press the retreating enemy. Sergeant Neal Emery led his platoon past the hills. Emery had just returned from a naval hospital on New Hebrides where he had been treated for yellow jaundice and malaria. "We went toward the ocean through really tall grass. As we started to clear the grass, a machine gun opened up and one of my men was killed, the last guy I lost. It was an American machine gun, too."[23]

The American perimeter was no longer in danger from enemy artillery. The men were sleeping on cots in tents, and while Japanese aircraft continued to fly night raids like the one that killed Clarence Bonderud, most men elected to stay above ground. This coupled with hot food prepared by Russell Opat and the other cooks led to a more relaxed atmosphere. Captain Albert Wiest and his machine gun and mortar men had been pulled back into XIV Corps reserve. His men, Wiest joked in a later reminiscence, were enjoying "some high spirits," by which he meant that they had made batches of "jungle juice" from raisins, dried peaches and other fruits. Others traded souvenirs or other items for medicinal alcohol or cash. "We would take Jap rifles and sell them to flyboys for forty-five dollars apiece," one later recalled. But Doug Burtell got his hands on something much better by sneaking into Colonel Moore's tent and taking two bottles of whiskey that Moore kept in a small box under his cot. "Moore was furious when he found it missing, but he never found out how it disappeared. It was good whiskey, too."[24]

One of the more bizarre incidents in these last weeks was an impromptu crocodile hunt that began late one night in the tent camp where Wiest had his men. "We furled the sides of out tents up at night, but we slept under mosquito bars. A couple of the guys were stirring up something out there. It was a crocodile, a big one too, about sixteen feet long. It was probably a female looking for a place to lay eggs. We were about a hundred yards from the Lunga River. It took off into the brush and they chased it. One of our guys was an Ozark hillbilly and he said 'I'll go get him!' Well, he did, pretty soon we heard some shooting." The battalion headquarters commander called Wiest to the phone and complained about the noise. "We had just a captain in charge right then, Colonel Hall hadn't come back yet. So he called and said 'what the hell is going on there?' I told him 'we got a crocodile.' He said 'you're crazy.' By that time, they'd dragged the crocodile back. I said 'leave that thing right in front of the company headquarters tent, the battalion commander's coming over.' He did and we had the evidence!" In the daytime the men hunted another pest, the giant crabs that would invade their tents. In grimmer times the crab's nocturnal noises had kept men alert for fear that Japanese infiltrators were moving about.[25]

1. The 164th Regimental Band in late 1940.
   (Courtesy of 164th Infantry Association)

2. Samuel Baglien, executive officer of the 164th Infantry, 1941.
   (Courtesy of Baglien family)

3. Douglas Burtell sketch of Company B roll call in Fargo. (Courtesy of Douglas Burtell)

4. Douglas Burtell at Camp Claiborne, Louisiana 1941. (Douglas Burtell)

5. Learning the finer points of the machine gun at Camp Claiborne. (Courtesy of Charlotte Engstrom)

6. Company D on parade. (164th Infantry Association)

7. John Paulson, Company F, at Claiborne. (Courtesy of Rudolph Edwardson)

8. Outside the Cow Palace in San Francisco, 1942. (Charlotte Engstrom)

9. Richard Stevens transferred into the 164th from the Kansas National Guard regiment. (Courtesy of Richard Stevens)

10. Damage from enemy battleship bombardment of Guadalcanal, October 1942. (Courtesy of U.S. Army Signal Corps)

11. John Kasberger (left) and Joe Horski, on Guadalcanal. (Courtesy of John Kasberger)

12. Ken Foubert, first man in 164th killed in action at Guadalcanal. (Courtesy of Foubert family)

13. Sketch made by a member of the 57th Engineers. (Courtesy of Alvin Tollefsrud)

14. Alvin Tollefsrud on Guadalcanal. (Alvin Tollefsrud)

15. Robert Hall, 3rd Battalion commander.
(164th Infantry Association)

16. Philip Engstrom of Company F. (Charlotte Engstrom)

17. Engstrom with Walter Rivinius at Kennedy Hospital in Memphis, Tennessee, 1943.
(Charlotte Engstrom)

18. Lt. Col. Arthur Timboe and wife on his wedding day, 1941.
(Courtesy of Mary Frann Timboe Riggs)

19. Japanese sniper nest on the Matanikau ridge line.
(Douglas Burtell)

20-21. Two "shoebox photographs" of Japanese prisoners and the U.S. Army cemetery on Guadalcanal, 1943. (Rudolph Edwardson)

22. Thomas Tracy, the "foxhole padre." (Courtesy of Fargo Catholic Diocese)

23. L-R: Joseph V. "Red" Meyers, Albert Wiest (Co M), Anton L. Beer (Co K), Kenneth Williams (Co L), Herald L. Crook. December '42. (Courtesy of Wiest)

24. Left to right: Ken Uthus, Odd Jacobson, Bob Olson, Robert Dodd, and Dave McCracken, all on Guadalcanal, the day before Christmas, 1942. (Courtesy of Robert Dodd, Jr.)

25. V-mail Christmas greetings from Clifford (Murphy) Fox, from Fiji, 1943. (Courtesy of Rita Fox)

26-27. Raymond Ellerman met Vicki Cooper on Fiji in 1943. The two married after the war. (Courtesy of Ray and Vicki Ellerman)

28. Army platoons practice with flamethrowers on Fiji, 1943. (Courtesy of George Isenberg)

29. George Dingledy, one of many replacements in the 164th. (Courtesy of Dingledy)

30. Rolf Slen (middle, back row) and his fellow crewmembers, part of the 494th Bomb Group in the Pacific in 1944. (Courtesy of Slen)

31. Emil Blomstrann (right) with friends John Wells and Joe Lauer, on Bougainville, 1944. (Courtesy of Blomstrann)

32. On the line at Hill 260, Bougainville, 1944. (Courtesy of William Dailey family)

33. Wounded on Guadalcanal in late 1942, Vincent (Swede) Clauson returned to D Company and fought at Bougainville and the Philippines. (Courtesy of Leatrice Clauson Cooper)

34. Clement (Murphy) Fox, Company D, who wanted to go back home and stay "for the rest of my natural life." (Rita Fox)

35. The "million dollar tree" on Bougainville.
(Douglas Burtell)

36. From Company E on Bougainville, left to right: Charles Ross, Charles Walker, Harry Mork, Pete Sherar.
(Courtesy of Charles Walker)

37. Lester Kerbaugh, highly decorated in the Pacific War, earned a battlefield commission in Korea. (Courtesy of Blake Kerbaugh)

38. A Company shower and laundry facilities on Bougainville, 1944. (William Dailey family)

39. Makeshift golf course in rear area, 1944. (164th Infantry Association)

40. Company bivouac on Leyte, 1945. (William Daily family)

41. Zane Jacobs, 1945. (Courtesy of Jacobs)

42. Joe Castagneto (left, on tree limb) serves his spaghetti supper to his buddies in the Philippines. (Courtesy of Castagneto)

43. Fred Drew in occupied Japan, 1945. (Courtesy of Drew)

44. John Tuff, postwar. (Courtesy of Edith Tuff)

45. Veterans of Company F bring Miles Shelley home to Carrington, 1947. (Courtesy of John Paulson)

16. Philip Engstrom and Walter Rivinius, 1999.
(Charlotte Engstrom)

47. Bernie and Mary Wagner, at 2008 regiment reunion.
(Courtesy of Bernie Wagner)

48. Bill Tucker and daughter, 2007 Association reunion.
(164th Infantry Association)

49. Joe Castagneto and Susan Tolliver-Pompa, at 164th annual reunion.
(164th Infantry Association)

50. For a fallen comrade. (Douglas Burtell Sketch, 1944)

On February 9, the lead patrols finished off some of the last defenders who were "in poor condition living among their own diseased and unburied dead." They then entered the deserted village of Tenaro at Cape Esperance. The advancing Americans passed discarded equipment, wrecked boats and other kinds of debris that told of a fleeing military force. In one spot they discovered tents of a Japanese hospital, with all the patients "on their cots dead." Some of the bodies were still warm. The Americans concluded that the wounded and sick men had been killed by injections of narcotics "to prevent capture when they could not be evacuated."[26]

Patch issued an announcement that the "total and complete defeat of Japanese forces on Guadalcanal effective 16:25 today." But that did not disguise the fact that the enemy had pulled off a remarkable escape. Admiral Nimitz later grudgingly admitted that "until the last minute it appeared [to us] that the Japanese were attempting a major reinforcement effort." The Japanese destroyer runs had evacuated about 10,000 men in an operation that had completely fooled the Americans.

"The news last night was terrific," Baglien recorded in his diary on February 10. "How happy us poor devils are. We have lived through 120 days of hell." There were still some Japanese on the island, meaning that no one was really prepared to discard the possibility of one final attack. Colonel Daley ordered that green signal flares be used to indicate an attack. If the flares were set off "along [the coast] it will be construed as enemy attempting landing." And so men remained on their guard. One accident occurred when a Company M man was wounded on the 12th -- a 60mm mortar crew in Company I fired a round at what they thought was a group of enemy stragglers. The wounded man survived. He was the last recorded casualty in the campaign.[27]

On February 21, Bob Hall, Frank Richards and Sam Baglien were all given orders "relieving them of duty with XIV Corps and to proceed by first available transportation to [the] continental United States." Baglien was ready to go. As his driver Gordon St. Claire remembered it years later, "Colonel Baglien was ordered back to the states . . . I remember the very words he used when he called, 'Slats (I was slim and trim at that time) would you come down and help me pack. I'm getting out of this rotten place!'"[28]

The three men did not leave before one final, solemn ceremony was performed on February 23, when the regiment mustered for dedicating the cemetery on the island. Harry Wiens recorded a nice account of this. "We all moved down into the perimeter and our units were drawn up in formation with the tall flagpole to our front, just forward of the cemetery. The size of our individual companies now resembled under-strength platoons. Father Tracy offered a brief, poignant prayer, with a moving tribute to the fellows we were leaving behind. Colonel Hall spoke for a longer time. He was his usual eloquent self, seeming to have recovered from his Matanikau wound. This was his farewell address, though of this we were not then aware." As the band played the Star-Spangled Banner and Taps, the men "came to attention and presented arms for the one and only time on this island." The ceremony over, the men fell out. Wiens and many of the others then went over to the graves "to say our final goodbyes to fellows we had known and shared with so much, friends who would live on now only in memory."

Kneeling by the grave of one of his friends, Wiens kept asking himself why these men died and he lived. It was a question many others asked as well. Art Nix had been evacuated to New Caledonia in late January with malaria and bronchitis. He regretted missing the cemetery ceremony and wondered why he survived. He had seen so much death in his job as a stretcher bearer, and realized that the individual was almost helpless in wartime. "I kept thinking 'what am I doing here?' I'm one person from a little town in North Dakota, one little speck over here. There was a certain amount of luck that determined your fate. A lot of it didn't make sense."[29]

Hall, Baglien, and Richards boarded a DC3 and left for New Caledonia soon after the ceremony. "Its goodbye to the 164th Infantry, may God bless them," wrote Baglien. All three would end up in California with training assignments. Bryant Moore had relieved the men he had deemed too sentimental. Baglien and the others equally believed that they were defending the regiment's traditions and preserving the men's morale. It was one of many irreconcilable clashes that occurred in the course of a long, bloody war.

The Lunga perimeter was rapidly being converted into a complex of bases from which the American forces would seize the rest of the Solomons chain. Over in New Guinea, MacArthur's 32nd Division had seized the Buna-Gona bases from the Japanese, also at great cost.

The 164th spent one more week on Guadalcanal. By now, the officers knew that the unit was to be shipped to the Fijis, and remain there for some time to "rest and reorganize." In something of a surprise, Paul Daley received orders to step down from command of the regiment, report back to XIV Corps, and take command of another combat unit. Ben Northridge became acting commander, but it was only a temporary assignment, as it was understood that a new commander would join the unit in the Fijis.[30]

As the day of departure approached, several men went out to visit some of the battle sites. Harry Wiens went up to Mount Austen to inspect the remains of the enemy entrenchments. Viewing them, he understood the losses the 132nd took in securing the summit. Other men revisited the foxholes where they held off the October assault. No one was interested in seeing Koli Point again. Gerry Sanderson spent time giving some minor medical care to island natives, who were now being paid by the Army to help expand the base facilities. As the time to leave approached, Sanderson thought back to when the Japanese sent over another air raid. He had run for cover and discovered his dugout filled with natives, "packed in, and sweating like something else." One of looked up at him and asked "Japaneee come?" Sanderson said, "yes, I'm afraid so." He wondered how it was that two civilized nations could turn a distant island filled with such simple folk into such a charred battleground.[31]

On March 1, the men boarded the troop transports *Hunter Liggett* and *Fuller*, bound for Fiji. As they boarded the ships, sailors were passing out candy and Hershey bars. Sergeant Wiens thought it was "ironically fitting" that here they were, men of war, munching on candy like kids as the ships up-anchored and left Ironbottom

Sound, where so many sailors and soldiers had died for possession of a jungle airfield. As Guadalcanal faded from sight, Wiens asked: "Could anyone who had not felt [what we did] know how bitter sweet the leaving really was?"[32]

(Endnotes)

[1] Frank, *Guadalcanal,* pp. 534-39; Toland, *The Rising Sun,* pp. 484-86; Miller, p. 337, for Japanese air power quote.

[2] The captured letter from Lance Corporal Koto Kiyoshi is filed with an edited copy of Samuel Baglien's diary in the 164th Infantry Association Papers at the University of North Dakota.

[3] The 147th Regiment, part of the 37th Division, had landed on Guadalcanal as part of the force intended to garrison the airfield that the U.S. Navy wanted to build east of Koli Point. When that plan was abandoned, the 147th joined the American forces in the Lunga perimeter. Most of the 2nd Marine Division was withdrawn from the island in December. Major General John Marston, the commander of the 2nd Marine Division, never joined his troops on Guadalcanal. Because by December there were more Army troops than marines, Frank notes (p. 556n), that "military punctilio" dictated that Marston, who outranked Patch, remain in New Zealand.

[4] Art Nix, tape given to James Fenelon. Christianson ended the war in command of the 27th Infantry. He and Nix remained in touch for the rest of their lives.

[5] Baglien Diary entries for January 1, 1943. In the post-war edited version that Baglien made available to the 164th Infantry Association, the last two sentences about the Moore promotion entry were not included.

[6] Colonel Russell 'Red' Reeder, *Born at Reveille* (New York: Duell, Sloan and Pearce, 1966), pp. 210-12.

[7] Richard Stevens, in his telephone interview with Shoptaugh, defended Moore: "He had just taken command, and had no real knowledge of the capabilities of his troops," and "the guard officers did not welcome him. He had to act decisively." A retired Marine officer, from the 2nd Marine Division told Shoptaugh that "it was obvious that some of the National Guard officers were not up to the challenge of commanding men in combat and had to be replaced."

[8] William Boyd, interview transcript, p. 20. An excellent introduction to the West Point influence is Theodore J. Crackel, *West Point: A Bicentennial History* (Manhattan, KS: University Press of Kansas, 2002).

[9] Robert Bruce Sligh, *The National Guard and National Defense: The Mobilization of the Guard in WWII* (New York: Praeger Press, 1992) show numerous examples of National Guard 'cronyism' (which when one reflects on it, is common in all professions). Ben Northridge was one exception to Baglien's prediction. Northridge was given command of the 3rd battalion while Hall was recovering from his wounds, and he forged a better working relationship with Moore than others had. One HQ man said he became known as "lying Ben" because he

cooperated with Moore. When the regiment left the island for Fiji, Northridge was given temporary command of it for the voyage. He was sent home before the unit went to Bougainville.

[10] Wiens, *My Little Corner of the War,* p. 138; Bernard Wagner, interview for Barnes County Historical Museum, April 20, 2000; William Welander interview with Shoptaugh.

[11] Wiens, p. 138-53.

[12] Miller, *Guadalcanal*, pp. 281-95; Robert C. Muehrcke, *Orchids in the Mud* (Chicago: J.S. Printing, 1985) pp. 134ff., for first-hand accounts of the 132nd Regiment's battles to secure the "Gifu" fortified positions. See as well Stanley Coleman Jersey, *Hell's Islands: The Untold Story of Guadalcanal* (College Station: Texas A & M University Press, 2008), pp. 350ff. The work of the natives on the island is summarized in the "Intelligence Annex to the Combat Experience Report, The Americal Division, 18 Nov. 42 to 9 Feb. 43," p. 18, copy in 164th Infantry Association Records.

[13] Brooks Kleber and Dale Birdsell, *The Chemical Warfare Service: Chemicals in Combat* (Washington D.C.: Department of the Army, 1966), p. 537. See also Leonard L. McKinney, *Portable Flamethrower Operations in World War II* (Washington D.C.: U.S. Government Printing Office, 1949) for some first-hand documents on the earliest uses of the flamethrower on Guadalcanal. Some men in the 164th, including Bernie Wagner, were trained in the use of the flamethrower while on Guadalcanal.

[14] Tony A, Hannel, "What a Helluva Night," *Guadalcanal Echoes*, November-December 1995.

[15] Welander interview with Shoptaugh.

[16] Gerald Sanderson interview with Shoptaugh, September 15, 2007, and Sanderson's November 21, 2005 interview for the State Historical Society for North Dakota SHSND).

[17] Wagner interview for Barnes County Historical Museum; Wiens, pp. 141, 147-48.

[18] Edwardson and Tollefsrud interviews with Shoptaugh.

[19] Sanderson interview with Shoptaugh; William Hagen interview transcript, pp. 32; Howard W. White interview by James Fenelon.

[20] M. E. Condon-Rall and Albert E. Cowdrey, *Medical Service in the War Against Japan* (Washington D.C.: Center for Military History, 1998), pp. 126-27.

[21] Dennis Cline and Bob Michel have written a fine first-hand narrative of the battle against malaria in their *Skeeter Beaters: Memories of the South Pacific* (2002).

[22] Cooper and Smith, *Citizen Soldiers*, p. 293; Miller, pp. 338-45; S1 Journal entries, January 25-30, for Savo Island patrol.

[23] Neal Emery interview with W. Michael Morrissey, June 14, 2005.

[24] Fred 'Fritz' Maier, taped remarks for Robert Dodd Jr., copy shared with Shoptaugh; Albert Wiest and Douglas Burtell, interviews with Shoptaugh.

[25] Albert Wiest and Doug Burtell, taped interviews with Shoptaugh, supplemented by Shoptaugh's telephone interview with Wiest, July 10, 2007; Cooper and Smith, pp. 292-93.

[26] Frank, pp. 596-97. The discovery of the dead Japanese in the abandoned hospital is recorded in the Americal Intelligence Annex, p. 17, 164th Infantry Association Records.

[27] Miller, pp. 348-49; Baglien diary entries for February 10 and 12, 1943; S1 Journal entry for February 10, 1943.

[28] S1 Journal entry for February 20, 1943; Gordon St. Claire, "The Wheel," *164th Infantry News*, June 1998.

[29] Weins, pp. 151-52; Art Nix, interview at Camp Grafton reunion [undated], North Dakota National Guard Historical Files, Bismarck.

[30] Baglien diary entry for February 22, 1943; S1 Journal entries for February 24 to March 1, 1943; Baglien diary entry for February 22, 1943.

[31] Sanderson interview with Shoptaugh.

[32] Wiens, p. 153.

Chapter 10: A Brief Respite
Fiji: 1943

On the way to Fiji, Bennie Thornberg wrote in his diary: "[Roy] Trimbo and I are sitting on the ship deck, not doing much of anything, just looking at the ocean. We are just going by the New Hebrides islands." As *Hunter Liggett* and *Fuller* made the five day voyage, the men performed light duties, cleaned up equipment, and carried out a couple of lifeboat drills. Mostly they slept. Meanwhile, on Guadalcanal, the hard-won island was being converted into a major base. Warehouses, machine shops, kitchens, supply dumps, and power plants sprouted like jungle vines. A completely new naval base was established at Koli Point. Trails became gravel roads and Henderson Field was expanded with more runways. When nurses arrived for the navy hospital, Seabees erected a movie theater.[1]

The Third Amphibious Force was headquartered at Koli, and it commenced its next step toward Rabaul. Halsey ordered the 43rd Division to occupy the Russell Islands and used it to build more fighter strips. He wanted to land troops on Bougainville, but the Joint Chiefs thought this was too risky without support from aircraft carriers, which were needed elsewhere. So Halsey made his next move against New Georgia.[2]

Happily, the 164th was free from involvement in these plans. As their ships entered Suva Harbor on March 4, the U.S. Navy gave them a fitting welcome, described by Elroy Greuel: "When we were a couple of hours from port, one of the ships took off ahead of us and we wondered where the hell he was going. Well, when we got to the harbor, that ship was at the entrance and all the sailors were in dress uniform, to salute us as we came in. That was really touching." The *Hunter Liggett* and *Fuller* could not dock at the crowded quay until the following day, but once the men debarked, they were ready to settle in for long rest.[3]

The British colony of Fiji was comprised of over two hundred islands, one of the largest being Viti Levu. Suva, the port at Viti Levu, was the colonial capital. Just over 4,000 Europeans lived in Fiji, mostly British. Indian and Chinese immigrants and native Fijians made up the rest of the 220,000 population. The battalions were trucked to camp sites around Suva. For a time, Company L was housed in an old school building and "in two grass shacks." But most men moved immediately into tidy wooden barracks built for British troops. Rudy Edwardson and his friend Perry McKechnie, using a handmade camera, took photos of the 2nd Battalion barracks, which looked like boxcars on raised platforms.

The Fijians, being of Melanesian and Polynesian ancestry, were just three generations removed from a near-stone age existence, but much had changed under British tutelage and the men of the 164th quickly came to like them. Fijian men were, as the British called them, "nice chaps." The women were friendly and attractive. Recalling this years later, one soldier noted that "several of us set out to discover just how friendly they could be." Some also discovered "a number of Hindi girls [who] plied the trade," including one madam who made donations to the local Red Cross office in an effort to keep the MPs from interfering with her business.[4]

The weather on Fiji was a wonderful change. The islands were free of the anopheles mosquitoes and malaria was all but unknown. This made little difference, however, because the malaria already in their bloodstreams soon ravaged scores of the soldiers, turning Gerry Sanderson into a busy ambulance driver. "We were making twelve, fifteen [hospital] trips a day; we'd get back [to the hospital] with one load and get word there'd be another one to take." Sanderson himself never had worse than a mild case. Most serious cases were treated by Atabrine and rest, but when Chuck Walker was admitted to hospital with malaria and yellow jaundice, he kept throwing up the Atabrine, forcing the doctors to give him quinine. Some patients simply had to be sent home. After his temperature reached 104°, Walt Byers was packed in ice and sent to New Zealand. Doctors decided that if he served in the tropics again, a recurrence would kill him. Home he went. Other serious cases of malaria went back to the states at intervals.[5]

Strangely there was a benefit from being hospitalized for malaria. American Army and Navy nurses were stationed at Fiji's hospitals -- and the patients got their full attention. After American nurses were captured by the Japanese in the Philippines during the early months of the war, nurses sent to the Pacific were not permitted to serve in combat zones. But they were sent to New Caledonia, Australia, New Zealand and Fiji during the summer of 1942. Violet (Vicki) Cooper, a pert Pennsylvanian who had joined the Women's Army Corps, was in Fiji. "We had one doctor, one dentist, us five nurses, and twenty corpsmen in our unit." In Suva, she dealt with patients who had malaria, fevers, and battle wounds. "You just felt so sorry for the men who you couldn't help. You just had to endure it." One evening Vicki went out with her friends to a post club to hear the newly arrived 164th Band. There she met Ray Ellerman. The truly fortunate Ellerman left Guadalcanal without a wound or even malaria. When he met Vicki, he was smitten with her. The two spent as much time together as they could. By the time Vicki left Fiji on a ship filled with patients bound for the States, Ellerman felt that this was the woman he wanted to marry.[6]

By early April 1943, the remainder of the Americal Division also left Guadalcanal and sailed to Viti Levu. While the ranks rested and reorganized, the Americal headquarters prepared its report on Guadalcanal. Officers from the Command and Staff School at Fort Leavenworth, earmarked for combat in the Pacific, came to Fiji to glean lessons about jungle warfare. Colonel 'Red' Reeder, Marshall's "eyes and ears" during his late 1942 visit to Guadalcanal, had his report published as a restricted booklet in mid-1943. Reeder's observations ran the gamut from weapons use to tactics and training.

In the main, the men interviewed by Reeder said they believed that their weapons were good but that the training they had received before going into a combat situation had been inadequate. Captain John Dawson (B Company), for example, urged "some [training] maneuvers on which men were deprived of food, water and comforts." From this a commander could learn which "NCOs and men can't take it." John Gossett, Company H, went further: "I have learned the primitive, rough and tumble way, [that] you can't pat all men on the back. You have to be rough with some men in order to get results." Lavern Lang, one of Gossett's platoon sergeants, simply said "for Pete's sake, teach the men not to be trigger happy," a sentiment echoed by Frank Richards who told Reeder that men had to be trained "in patience . . . wait for the enemy to expose himself." Sergeant Clair Arrowood, who had seen as much combat as anyone with Company F, probably gave the best advice for future combatants when he said "my message for you to take back, Colonel, is to stress *real* scouting and patrolling and to teach them to go *the hard way.*" All of those interviewed agreed that the enemy would fight to the death.[7]

The Americal Division's own report mirrored the observations in Reeder's report: "The morale of the Japanese soldier, even under adverse conditions, is good. This may be attributed to their high-mindedness, amazing zeal, and a feeling that sacrifice and death for the emperor is the highest aim in life." The report noted that every one of the barely two hundred prisoners taken by the division was "anxious that their name[s] would be concealed in order that their relatives at home would not be persecuted." In short, those who came to the Pacific theatre in the months ahead could expect a long, bloody war.[8]

The regiment's role in the seven month struggle for Guadalcanal was largely unknown to the American public. The first time that a major national press publication mentioned the performance of Army troops on the island was a month after the Henderson Field fight in mid-November, when *Time* magazine credited "green but eager" GIs for clearing out the Koli Point salient. While the Marine Corps public relations officers in Washington carried out a well-organized publicity blitz, Marine releases did not mention the Army's role until after the island was secured.[9]

Some newspapers in North Dakota had revealed in late October that the regiment was at Guadalcanal. Major Harry Tenborg had written to his wife on October 18, revealing that he was at Guadalcanal, describing his unit's condition and asking his neighbors in Carrington to "pray for us." Other letters from men in the 164th were received and published soon after this, also noting the Guadalcanal location. On December 10, 1942, Rilie Morgan Sr., the editor of the *Walsh County Record* published a moving obituary of his son, and referred to the letters his son had written "from Guadalcanal."

It was not until February 11, two days after the island was secured, that Secretary of War Henry Stimson issued a statement that officially identified the 164th as being on Guadalcanal. The Fargo *Forum* published this with the comment that the 164th's location "has long been common knowledge," but that most papers had refrained from saying so in deference to military censorship.[10]

The December 1942 issue of the *Infantry Journal* printed a few photographs of soldiers with short text stating that, "shortly after the Marines landed," Army troops from a "Solomon Islands Task Force" had sailed from New Caledonia to reinforce them. One photograph identified "Brigadier General E. B. Sebree and Colonel Bryant E. Moore," saluting men boarding a ship and stated these officers had "reviewed" the troops just prior to their departure. Foster Hailey wrote a series of articles in the *New York Times*, based on his visit to the island in February. Published in May 1943, Hailey's series praised the regiment. His May 28 story, "164th Won Glory at Coffin Corner," was based on interviews he did with several men in Company E. William Hipple, an

Associated Press correspondent, also visited the island in February and talked to several men and then wrote a story crediting the 164th for preventing the loss of Henderson Field in the "bitter fighting" of mid-October.[11]

Among those who had served with the 164th, Chesty Puller and General Vandegrift offered public praise. Puller, home recovering from his wounds, told reporters in January 1943 that the "army unit" was "a very good force indeed and I would be happy to command more like them." Vandegrift's unit commendation was published soon after, and when Vandegrift returned to the states, he told North Dakota officials that the unit "made as fine a body of soldiers as were ever in the army, and that he would be glad to welcome the entire group complete into the Marine Corps." Meanwhile, General Patch wrote to General Marshall that "the Army [has] made a larger contribution to the operations here on this island than is properly understood." He asked Marshall to encourage more press coverage of the Army in the Pacific.[12]

North Dakota papers devoted more coverage to the regiment after officers and wounded men returned from the Pacific. Sam Baglien returned to the state and granted several interviews drawing upon the contents of his diary. The Fargo *Forum*, the state's largest newspaper, drew on this for a five-part series in late April, which culminated in a list of the men who had been killed in the fighting. Arthur Timboe a few weeks later told a local reporter that he "preferred to talk but little about that South Pacific nightmare," although he was proud of "the courage and fighting ability of his men." He spent the rest of his trip with his wife and his daughter Mary, born just before he went to Guadalcanal.[13]

John Hagen's father wrote to him about this publicity. In response to Dr. Hagen's news that two of his friends back home were now in the service and becoming "anxious for a little action," John replied as a veteran: "Well, experience is anyone's teacher. They will change their minds, I believe."[14]

Fiji seemed like the ideal place to unwind. Doug Burtell recalled that he and Bob Dodd walked to Suva once and spent an afternoon enjoying themselves at one of the bars. As the evening wore on, they started back, "pretty loaded." Dodd suddenly said he was too tired to walk and the two men "borrowed" an Air Force jeep left on the road. "We caught hell for that and they took away our stripes. As soon as we got ready to go to Bougainville, we got our stripes back." Rudy Edwardson remembered that many mornings one or more men, hung over, hid underneath the barracks while someone answered role call for them. At a dance arranged one night, many of the men were horribly drunk. When Charles Walker upbraided the local MPs for not controlling the men, he saw that they were inebriated, too. Officers during this time recognized that the war would resume soon and forgave all but the most serious transgressions. Alcohol was plentiful but female companionship more problematic. Most enlisted men were warned off of flirting with American nurses, "reserved for officers." Fijian women were friendly, however, although some men were intimidated by the Fijians' forwardness and athleticism. Chuck Walker recounted an impromptu field hockey game between soldiers and Fijian girls who "had walked and run all of their lives . . . They played the men out in a matter of minutes, running like deer."[15]

Promotions and transfers rapidly transformed the regiment. After recovering from malaria, Alvin Tollefsrud went into the divisional headquarters as part of the post office detail. John Hagen received a transfer to the Army Air Corps and left for the States to become a pilot. John Stannard, now recovered from his wound, was also returning to the U.S. Because of his success at Guadalcanal, Stannard was one of two men from the Americal Division to receive an appointment to West Point. At least four of the men that had fought alongside Stannard received battlefield commissions. Lawrence Poe was promoted to First Sergeant for E Company, and then received a transfer to the Air Corps, where he became an air gunner. Others went off to other units.[16]

A number of veteran sergeants and men volunteered for a special mission in Burma. Les Aldrich, one of these men, later explained to an interviewer that "we had been offered a deal to do this one mission and then they said we could go home. When we got there we were part of the Merrill's Marauders operation and we ended up being out there a lot longer than we expected."[17]

Like Walt Byers, Bennie Thornberg left the Pacific due to malaria. When they told him he was headed "to the states," he wrote "sure sounds good" in his diary. Bill Welander was similarly afflicted by malaria. Despite having consumed enough Atabrine to make him "so yellow I made a Japanese look pale," he was still quite debilitated and the medical staff determined that he, too, should go to McCloskey General Hospital in Texas where he was "a guinea pig" for testing malaria treatments. There he took upwards of thirty-five Atabrine tablets a day and other, more experimental drugs. "I can still swallow a handful of pills with no problem," he admitted sixty years later. Neal Emery, who had never recovered properly from yellow jaundice, was also sent home. He spent "weeks and weeks" in a military hospital in Topeka, Kansas. The Army wanted to assign him to train men for jungle warfare but his doctor insisted his malaria was too likely to flare up again and kill him if he was in a warm climate. So he was discharged late in 1943.[18]

Vernet Anderson, a rifleman with C Company, was down with malaria and problems with jungle rot. He had applied for OCS training, but his weakened condition led the doctor to give him a full physical and decided that, with the malaria and trouble from a pre-service operation he had as a kid, he, too, should go home. Anderson left Fiji with mixed feelings, happy to go but thinking also of those who stayed behind: the dead who "had fought a good war," and those in his unit who had kept him alive on Guadalcanal, especially "Mr. Company C, our first sergeant Joe Wangness [who] helped me and the others so very much," . . . Arvid Honsvall, another platoon sergeant, and Joe Bergh, a cook who spent many nights with us [at Matanikau] when we were short of people. We had a lot of fine officers during this time, we learned from them and they from us."[19]

The leadership that Anderson credited for his survival was also changing. At the top, Major General John R. Hodge took over command of the Americal Division. Hodge was not a West Point man; he had received his commission in 1917 through the OCS program and had fought in some of the bloodiest battles of 1918. Before coming to the Americal, he was the assistant division commander of the 25th Division on Guadalcanal and commanded the 43rd Division during the last stages of the New Georgia campaign. An instinctive fighter, Hodge favored the all-out offensive as the means to win. By the end of the war he was being called "the Patton of the Pacific" by correspondents. The new 164th commander was Crump Garvin, a West Point graduate. He had served in administrative positions so far in the war, as chief of staff for the XIV Corps and then as chief of staff for the Americal. When Hodge offered him a regiment, he chose the 164th because Hodge thought it "the best regiment" in the division. Like the departed Bryant Moore, Garvin had no combat experience prior to Guadalcanal; unlike Moore he generally got along with his headquarters staff and with the battalion and company commanders. Among Garvin's staff was Fred Flo, promoted to head of the S2 section.[20]

New Battalion commanders were appointed to replace the departed Hall, Timboe and Richards. At least two of these men were old Guard officers: Bill Considine was made commander of the 1st Battalion, while Stafford Ordahl took up the mantle of the much-admired Bob Hall. Lt. Colonel William Smith was made commander of the 2nd Battalion, but for unknown reasons he was soon replaced by a man named Floyd Dunn. Charles Walker developed respect for Colonel Dunn: "He had common sense, was a real square shooter," and "was a nut on training." When Dunn learned that Walker had collected a cache of Japanese weapons and was testing them, "he called me to headquarters to ask why the battalion [staff] hadn't been invited to attend the [weapons] demonstration." The upshot of this was an extensive series of training maneuvers using Japanese weapons and tactics, and Walker's promotion to Captain as the battalion's S3 (plans and training) officer.[21]

Other men advanced in rank included John Tuff, who was promoted to the rank of lieutenant and put in charge of training exercises for Company M. Several freshly minted second lieutenants transferred in from stateside OCS programs. Rough around the edges, most were eager and earnest, but there was one rat who Al Wiest and Walker caught cheating at poker.[22]

By the end of March replacements were coming in to fill out the ranks. They were sorely needed. Deaths, wounds, illness, and transfers had depleted the rank and file which meant that less than half of the 2817 men who had landed on Guadalcanal were still with the 164th. The veterans wondered just how well their replacements would fit in, but the wisest vets, realizing that they, too, had been rookies, made efforts to reassure the new men. Sergeant John Paulson told the replacements under his command that "we're in this together. When we go on a mission, I'll explain it to you and you can ask any questions. I'm no better than you are. We've got to work together as a team."[23]

Replacements came from all parts of the American map. Two men, Joe Castagneto and Gene Scolavino, were among those who now joined the regiment. Castagneto was a Massachusetts native who had left high school at fifteen and worked in a machine shop before he enlisted in the army in 1942. He was training with a new regiment in 1943, "qualified as a sharpshooter" and was proficient with the Browning Automatic Rifle. "They made me the assistant squad leader. Almost all of our training was for Europe and we thought we would soon be sent there." But as the Pacific needed new BAR men, Joe's commanding officer called him in and said he was heading west, not east. "He said, 'I got a letter here from General MacArthur, he's been looking for you,' that's how he told me." Castagneto was soon on an "an old tub of a ship" heading for the South Pacific. He spent most of the voyage acting as a lookout in the ship's crow's nest: "that was awful, swaying up there day after day looking for a sub's periscope."

Eugene Scolavino, a year younger than Castagneto, was from Providence, Rhode Island. His dad had died when Gene was about a year old. "So I left school early, too, like Joe, and worked at several jobs, running a multigraph machine at a whole-sale grocery. I tried to join the Navy in 1942, but I had only 20/400 vision in my right eye so they rejected me. I was still working in 1943 when the Army sent me greetings." Short and slight of build, Gene was trained as a medic in an eight week course, and then volunteered for overseas duty. He met Castagneto on the ship.[24]

Both men were assigned to Company A's 1st platoon on Fiji. Castagneto felt warmly welcomed. "They were glad to see us, they treated us very well, because they knew there was going to be another fight and they needed us." Scolavino's welcome was a bit different. "I was surprised to hear them call out '164th Infantry' when they assigned me. I weighed about 120 pounds, and was so short the uniform didn't fit me too well. I'd never shaved yet. And I'll never forget it, when we got to Fiji, the sergeant who handled the company mail saw me and he was a big guy. He looked at me and asked, 'you're not a soldier, are you?' I said, 'yeah I was.' He turned to another guy and said '*are we losing this war?*' He thought they must be drafting kids at home."

The Americal Division issued a series of training orders to all squads, calling on the trainers to "emphasize the 5 Fs" – "Finding, Fixing, Fighting, Finishing and Fending [off counterattacks]." Once scouts found an enemy force, "the enemy must be fixed in place" with fire "to prevent him from running away." Using fire and maneuver, the enemy could be "finished off" but the attackers must watch their "front, flanks and rear at all times" in order to fend off counterattacks. Flanks had been neglected often enough on Guadalcanal for the commanders to demand that this be a matter of "constant application" in all "practical training and operation." Every officer, no matter how much experience he had, was drilled on this. New officers were ordered to "control, direct and dominate [their] men under any situation," and to remember that only hard work and "taking care of your command" would win "the complete loyalty of your men." As will be seen, while many officers took this to heart, some did not.[25]

The Americans arranged to train with Fijians that the British colonial government organized as commandos for scouting and harassing Japanese units. Fijians proved to be ideal warriors. They were in superb condition, able to march for hours with limited rest and little sustenance. Much more at home in the jungle terrain than either the Americans or the Japanese, they had the ability to blend into the terrain, approach any man in complete silence and come within an arm's length of him without being discovered. On Fiji, the scouts played a chilling little game with the Americans. After being warned repeatedly to post guards and remain alert, platoons on bivouac would bed down only to awake the next morning to find equipment laid out around them, and chalk marks on their shirts, making them 'dead.' No man had heard a sound, no night guard had seen a thing, while the natives had gone soundlessly though the camp. The Fijians returned later with a large sack, and Doug Burtell watched as they pulled out it "helmets, billfolds and watches that they had taken from the sleeping men." Charles Ross, a young lieutenant newly attached to E Company, went on practice reconnaissance patrols with Fijian scouts. "Watching them, I was glad to know they were on our side." The Second Fiji Commando was earmarked to go to Bougainville.[26]

While the regiment was being rebuilt the war went on. In the European theater, the Allies drove the Germans and Italians out of North Africa and then invaded Sicily. American and British bombers were now bombing Hitler's industries and cities. The Russians forced the surrender of a German army at Stalingrad. Roosevelt met Churchill and Stalin at Teheran to coordinate plans for the Anglo-American invasion of the French coast. The initiative in Europe now lay with the Allied powers. The pendulum had swung in the Pacific as well. Using decoded messages, Army fighters ambushed Admiral Yamamoto's plane and killed the mastermind of the Pearl Harbor attack. MacArthur continued his advances in New Guinea, while the Japanese lost hundreds of aircraft and irreplaceable pilots trying to stop him. In 1942, American forces had reacted to the enemy; now it was the U.S. that was dictating the direction of the war.[27]

In November 1943, the 3rd Marine Division landed on Bougainville and established a perimeter between Empress Augusta Bay and the Torokina River. American air power on Bougainville could directly threaten the enemy's base at Rabaul. If Rabaul was neutralized, the way to the Philippines was possible, especially with the revived Pacific fleet: six new aircraft carriers, each carrying as many as a hundred aircraft. The *Essex*, the first of this new carrier class, was commissioned on the last day of 1942 and five more had since joined it. Organized into task groups, the "fast carriers" became the key to the strategy that Admiral Nimitz and his staff had devised for carrying out a "two track" thrust into the central Pacific -- one, under Halsey, up the Solomons, the other toward the Mariana Islands from Pearl Harbor. From the Marianas heavy bombers could reach Japan.[28]

Along with the carriers, a host of "little ships" were now available. Foremost among these was the LST (landing ship tank, but more usually called by its crews a "large slow target"). Capable of carrying men directly to a beach, over a thousand of these were built during the war. New destroyers and cruisers, new speedy battleships able to keep up with the carriers, and hundreds of new supply ships rounded out a growing naval force that Japanese industry could not hope to match.

American land-based air power was much greater as well. The Guadalcanal perimeter had been defended for months by a handful of the stubby Wildcats. Now, the Solomon Islands Air Force (Airsols) was operating out of Guadalcanal, New Georgia and two new airfields on Bougainville. The B-17s that had been on Guadalcanal had been replaced by squadrons of B-24s, which could carry the same bomb load as the B-17, but carry it further, an important advantage in the vast Pacific.[29]

One of the many marvels achieved by the armed forces in this conflict was finding and then training the thousands upon thousands of pilots and air crew for the expanded air forces. One of these was Rolf Slen. Looking back on it later, Slen recalled that he had no idea where Pearl Harbor was when he heard it was attacked. "I was a 17 year old high school senior [in Minnesota]. We had a Luther League meeting at church that Sunday night. Dorothy Nickolai, a classmate of my brother, played a selection on the piano, entitled 'Japanese Lullaby.' "She apologized for it and said she should have practiced for and played something else."[30]

After leaving high school, Slen attended college and then, aware that the draft board would inevitably come calling, applied to the Army Air Force's "air cadets" program. He was accepted and began his cadet training in Kansas City. In the end, he "wasn't very good" at piloting and washed out. "They decided to classify me for navigation school instead." After navigation school, Slen joined a B-24 crew as "a freshly minted Second Lieutenant . . . 19 years old and looked younger." Months of training flights followed and then he was assigned to a group that received orders for the Pacific. As he plotted the route out of San Francisco, he wondered if he would "live to see the good old U.S.A. again." Slen's duties and those of the 164th would cross many times in the months ahead.

Each new island airbase brought the American forces closer to the Japanese home islands. It was the job of the ground troops to secure and defend enough real estate on which the airbases could be built. With air power it was not necessary to wrest every base from the enemy. But with the fanaticism of the Japanese soldiers, it seemed to the ordinary American foot soldier that he would have to kill every one that he encountered. The GI of late 1943 was better armed and equipped than he had been in 1942. The uniforms were lighter and better adapted to jungle terrain and climate. More automatic weapons had been added to increase a squad's firepower. Attacking forces now used "corkscrew" tactics, with flamethrowers and improved explosives to tackle the toughened fire pits the Japanese were so adept at building.

Two divisions, the Third Marine and the 37th Infantry, had landed at Cape Torokina on Bougainville's southern shores. Empress Augusta Bay, extending outward from the Cape, provided ample room for naval support, and the Americans quickly built airfields and a permanent base. Over a hundred and thirty miles long, Bougainville was much like Guadalcanal, with a central spine of high, rugged mountains, including two active volcanoes, tropical rain forest over hilly surface, and a shallow rapidly running river that ended in swamp. The air was alive with mosquitoes and flies. The island was rife with the familiar malaria and dengue, but a dangerous form of scrub typhus, carried by lice, was also present. The rainfall and molds could reduce leather boots to mush in a matter of days.

The island was populated by about 40,000 natives scattered into fairly small villages along the coast. Because there were several German missionaries on the island, many natives were not especially friendly to the cause of the Allies. Some began to change their minds once the Japanese took control. A formidable force of 60,000 Japanese was stationed on Bougainville, mostly concentrated largely in the southeastern end of the island at Buin and on the offshore Shortlands Islands. Another group was stationed in the north around the Buka Passage. In order to shield Rabaul the enemy meant to hold it to the last man.

The Americans did not have very reliable maps or detailed information about Bougainville. Once again, the perimeter was surrounded by hills ranged about the Bay. That meant that just like Guadalcanal, the Japanese inevitably would send troops to counterattack the American lodgment. Admiral Halsey and his planners had to accept this. It would have taken a total of six or more divisions to take control of the whole island, and Halsey had no such force. As the naval historian Samuel Eliot Morison summarized the probable situation, "it would have been folly to send men out looking for trouble." Instead, "by fortifying the Perimeter and letting the enemy attack him, [the American commander] had the choice of weapons and terrain, a wide-open sea entrance and superiority in strength at point of contact." Halsey wanted to use his Marine division for a new assault, so the Americal was selected to join the perimeter.[31]

The men of the Americal were alerted on November 25 that they would go to Bougainville. They were as ready as a rebuilt regiment could be; the unit's history commented that training on Fiji had been "excellent . . . generally performance by the old men was better, due to more hardened condition, acclimation to the tropics, and [battle] experience." The 164th was chosen to be the first Americal regiment to land, and it sailed from Viti Levu on December 19. Rookies in the ranks and even some veterans thought they might be bombed and shelled after landing. Over in Company L, men recalled one of their buddies who had been "deathly ill" with malaria for some weeks and upon release had been "warned to take it easy and abstain from happy water." Ignoring the docs, he dashed out to the local bars, swam in the river and took up "with a dusky maiden." Fevered again with malaria, he was sent back to the states. One of his friends later admitted that "many [others] tried evening swims, but they were all sent to Bougainville just the same."[32]

Arriving at Empress Bay on Christmas Day, they were surprised to see no Japanese aircraft in sight. They landed in some rough water, but that was about all of the discomfort. A carrier strike on the harbor at Rabaul had damaged several Japanese warships, ending threats to shell the beachhead while destroying dozens of enemy planes and inflicting more structural damage than the Japanese could repair. The regiment moved forward, crossed the Piva River, and filed into the perimeter entrenchments of the departing 9th Marine Regiment. The companies had the rest of the marines on their right and the 37th Division on their left. The 182nd Regiment with attached artillery was expected to land in three or four days. The 132nd's arrival would complete the division's roster in early January. So far, Corps headquarters told them, there were signs of Japanese patrols, some sniping, but no indications of a major enemy attack.[33]

New Year's Day was a week away. Maybe this assignment would be okay, they said to one another, maybe 1944 would be a good year.

(Endnotes)

[1] Ben Thornberg diary entry for March 3, 1943. Roy Trimbo was not related to Arthur Timboe, who was still on Guadalcanal with XIV Corps Headquarters. The transformation of Guadalcanal is well described in Samuel Eliot Morison, *Breaking the Bismarcks Barrier* (New York: Little Brown, 1950), pp. 100-103.

[2] Morison, pp. 3-22; Edwin Hoyt, *How they Won the War in the Pacific* (New York: Weybright and Talley, 1970), pp. 210-218. MacArthur had by then seized Buna and wanted to advance far up the coast of New Guinea. But the strategy of by-passing enemy bases, to 'wither on the vine' could not be implemented until more long-range bombers reached the South Pacific.

[3] Elroy Greuel with Shoptaugh.

[4] Rottman, *World War II Pacific Island Guide*, pp. 92-97; Thurston Nelson, "From Prairie to Palm Trees" (on housing of L Company and local prostitution).

[5] Edwardson and Sanderson interviews with Shoptaugh; Walt Byers interview at the National Guard headquarters in Bismarck; Charles Walker, *Combat Officer*, p. 70.

[6] Ray and Violet (Vicki) Ellerman, telephone interview with Shoptaugh, September 19, 2007. Vicki Cooper's voyage back to the States was an adventure all its own: the freighter sued as a hospital vessel broke down after sailing through a storm. It's propeller damaged, it drifted for several days before another American ship arrived to tow it into a port for repairs. "It was so hot below decks that we brought the patients up on the main deck for fresh air," Cooper recalled. "One man was deranged by the heat and his fever, and he jumped overboard. We couldn't rescue him." Cooper lost so much weight during her Fiji assignment that the Army gave her stateside duties for the rest of the war. She and Ellerman married after he came home from the Philippines in 1945.

[7] *Fighting on Guadalcanal* (U.S. Government Printing Office, 1943) pp. 50-57. Reeder's remarks about his conversation with Bryant Moore, regarding the leadership problems in the regiment, were not included in this report.

[8] "Intelligence Annex to the Combat Experience Report, The Americal Division, 18 Nov. 42 to 9 Feb. 43," Tabs B and E, both documents in 164th Infantry Association Papers. Some of the men interviewed for this study recalled that while on Fiji they met with representatives of the Springfield armory for details on the performance of the M1.

[9] "Battle of the Pacific," *Time*, November 16, 1942. The same issue praised press correspondents on Guadalcanal for being "As Tough as Marines," but this second article made no mention of U.S. Army forces.

[10] "My Son," the pamphlet version of Rilie Morgan's obituary, was provided to me by Todd Morgan, current publisher of the Walsh County newspaper. See also "Relatives and Friends will be Proud of Record," Fargo *Forum*, February 11, 1943.

[11] *Infantry Journal*, December 1942, p. 5, accompanying article on "Xmas Present." Portions of Hailey's stories later appeared in his book *Pacific Battle Line* (New York: MacMillan Co., 1944). Hipple's AP story appeared in several newspapers. See also "Devil's Lake Man Back from 'Guadal'," *Devils Lake Daily Journal*, May 1943.

[12] Puller's remarks are from story "Officer Pays Tribute to ND Troops," Fargo *Forum* [January 1943]. Wyman, *Sandy Patch*, p. 65, for the letter to Marshall. The public perception of Guadalcanal was much determined by Richard Tregaskis's book *Guadalcanal Diary* which became a best seller and the basis of a major motion picture. He left the island three weeks before the 164th arrived.

[13] Fargo *Forum*, April 21-25, 1943 for series. Baglien later expanded upon this interview and used his diary to publish his account of "The Second Battle of Henderson Field," in the *Infantry Journal*, May 1944.

[14] John Hagen to Olaf Hagen, March 17, 1943, Hagen Family Papers, NMHC.

[15] Burtell interview with Shoptaugh; Edwardson conversation with Shoptaugh; Walker, pp. 79-80.

[16] Tollefsrud interview with Shoptaugh; John Stannard, *Battle of Coffin Corner*, p. 116. Stannard preserved a detailed account of his being wounded, medical treatments and appointment to West Point in a journal he wrote while in the hospital. Blake Kerbaugh, who obtained access to this journal, kindly provided me with a copy.

[17] Aldrich interview with Mary Helm.

[18] Ben Thornberg Diary enetries for August 9, 13, and 29, 1943; Welander interview with Shoptaugh; Neal Emery interview for Barnes County Historical Museum.

[19] Vernet Anderson memoir in *A 100 year Look at Grafton, North Dakota, 1882-1982* (Grafton: Centennial Corporation of Grafton, ND, 1982), pp. 386-87.

[20] Hodge and Garvin from Cooper and Smith, *Citizens as Soldiers*, p. 329. Ben Northridge, the short-term commander of the regiment, was assigned to a training base in Arizona, where he offered a ride to two soldiers who he did not know were deserting. They assaulted him and dumped him in the desert. He survived but with disabling brain damage.

[21] "History of the 164th Infantry, July 1-1943-September 30, 1943, p. 102, in 164th Infantry Association Records; Walker, pp. 72-79, and Walker interview with Shoptaugh, January 28, 2009.

[22] "Account of Enemy Operations on Guadalcanal" in John Tuff files (courtesy of Edith Tuff); "History of the 164th Infantry, July 1-1943-September 30, 1943," p. 102, 164th Infantry Association Records.

[23] Paulson interview with Shoptaugh.

[24] Here and below, Joseph Castagneto and Eugene Scolavino, joint interview with Shoptaugh, September 2007.

[25] Americal Division Headquarters, "Training Memorandum No. 12," 29 July 1943, preserved in Frank Doe Papers. A second copy of this memorandum, together with the memorandum "Responsibilities of Commissioned Officers," 13 November 1943, is in the papers of John Tuff (courtesy of Edith Tuff).

[26] Colin Larsen, *Pacific Commandos: A History of Southern Independent Commando and First Commando Fiji Guerillas* (Wellington, NZ: A.H. and A.W. Reed, 1946); Burtell interview with Shoptaugh; Major General (ret.) Charles Ross, conversation with Shoptaugh, September 16, 2008.

[27] MacArthur's campaign is well delineated in two books by Harry Gailey: *MacArthur Strikes Back: Decision at Buna* (Novato, CA: Presidio Press, 2000), and *MacArthur's Victory: The War in New Guinea* (Novato, CA: Presidio Press, 2004).

[28] Clark Reynolds, *The Fast Carriers: The Forging of an Air Navy* (New York: McGraw-Hill, 1968) is excellent for the new carrier fleet. When these were devoted in November 1943 to the assault on Tarawa, Halsey's staff complained that Bougainville was being treated as a "shoestring" affair. Nimitz did temporarily detail one carrier task group (CV *Essex* and CVLs *Independence* and *Bunker Hill*) to suppress enemy air power on Rabaul. This action was very important to the success of the landings at Cape Torokina.

[29] Eric Bergerud, *Fire in the Sky: Air Power in the South Pacific* (New York: Basic Books, 2001) is very good on the evolution of land-based air.

[30] Here and below, Rolf Slen, "A Memoir of World War II," (privately printed, 1997), and Slen interview with Shoptaugh, July 29, 2007.

[31] Morison, *Breaking the Bismarcks Barrier*, pp. 280-304, 426; Harry Gailey, *Bougainville, 1943-1945: The Forgotten Campaign* (Lexington, KY: The University Press of Kentucky, 1991), pp. 22-38, 132-33. (Gailey's detailed study updates and expands upon the Bougainville chapters in John Miller's official Army history, *Cartwheel: The Reduction of Rabaul* (Washington DC: Office of the Chief of Military History, 1959).

[32] "History of the 164th Infantry, July 1, 1943-September 30, 1943, p. 97-98, in 164th Infantry Association Records; Nelson, "From Prairie to Palm Trees."

[33] United States Strategic Bombing Survey (Pacific), *The Allied Campaign Against Rabaul* (Washington DC: Naval Analysis Division, 1946), pp. 25-32; Gailey, pp. 78-92, 132-33; S2 Journal entries for December 25-27, 1943.

Chapter 11: Perimeter Defense on Bougainville
December 1943-March 1944

Landing on Bougainville on Christmas had its benefits. Dick Stevens fondly recalled "our three Christmas dinners." There was one on Fiji before they left and a second on board ship the morning before they landed. "We were the best-fed debarkees in the Army," when they landed and made camp, only to discover that the supplies brought ashore included a third dinner, "fresh-cooked turkey, no longer more than slightly warm, but fresh turkey." It was certainly a finer welcome than the one he had received the year before at Guadalcanal.[1]

There was virtually no fighting going on. Combat engineers and Seabees had constructed strong fortifications on the perimeter, looking out over "fields of fire . . . for 100 yards in front of the lines." Beyond this there were advance warning posts. Aircraft flying from the two airfields suppressed snipers and kept an eye on the jungle trails to the north and east. Patrols out past the perimeter located signs only of small numbers of Japanese. An exception occurred when B Company sent out a large patrol of two officers and fifty riflemen toward the Torokina River, where they found and ambushed five enemy soldiers. During the return march though, additional Japanese, perhaps drawn to the shooting, began to snipe from the brush. A running fire fight ensued in which Sergeant Thomas Clewes was killed and two other men were wounded. The patrol report emphasized that the Japanese seen appeared in good health and well equipped. This was one of the first signs that fresh enemy troops were moving into the area. Additional patrols made maps of key hilltops, including Hill 600, site of a future battle. Concerned by the signs of an enemy buildup, Major General Oscar Griswold, the commander of XIV Corps, extended his northern front. He ordered Colonel Garvin to advance 1000 yards past the perimeter and take one of the hills. Garvin used part of the 3rd Battalion for this chore. The dense jungle they encountered proved to be more of an obstacle than the enemy, who fought from a few machine gun nests that the men reduced with artillery and grenades. American casualties were light. Once the summit of the hill was secured, the ground was turned over to the 37th Division.[2]

The most serious combat for the Americal during the early weeks came in late January and involved the 132nd regiment. With Empress Augusta Bay filled to overflowing with ships and landing craft, all engaged in bringing to shore the thousands of tons of supplies, Griswold did not want a sudden enemy thrust from the Torokina River to jeopardize his beach. He therefore sent the 132nd's G and F Companies across the river to destroy any enemy concentrations. The troops commenced their attack on January 20, moving forward with the support of flamethrowers and Sherman tanks to clean out enemy nests. As usual, the Japanese had carefully disguised their pillboxes. It took reinforcements and two weeks of hard fighting to get the job done. A sergeant from B Company, 132nd, Jessie Drowley, directed a Sherman tank toward positions he had spotted. Riding atop the tank Drowley pointed out one nest after another. He was wounded in the chest and temporarily blinded by another wound to his face. But he stuck it out until the nests were destroyed and lived to get the Medal of Honor. The "Hornet's Nest," as the troops had called the position, was eliminated on February 6.[3]

Griswold had about 45,000 men under his command. He accepted that the Japanese would eventually make an all-out attack on the American positions. So his engineers turned the perimeter line into a rock-hard stronghold while others laid out interior roads to allow units to move quickly to any threatened sector. Willard Garris, a Navy Chief Petty Officer assigned to work on these defenses, wrote to friends that "it is complete, though then what [I will do] I couldn't say. I suppose we will be moving on" which he regretted because Bougainville was so far pretty pleasant duty. "We really haven't had it so bad after all . . . We are becoming quite civilized. A [picture] show every other night and sometimes a little home talent the other nights. The U.S.O. is putting on for us tonight. We stayed for a little while and finally gave up as it is raining too hard." He assured his friends that although the Japanese staged occasional night air raids, he was not in much danger.[4]

Joe Castagneto's first duties on Bougainville were to build stairs and furniture. "We got up to the front where A Company was going to be, and we were at the foot of a steep hill up where our foxholes were. Our captain told me and a bunch of others that we needed to build some steps up the hillside. So I got busy cutting big stands of bamboo" for framing terraced steps. "We used bamboo for everything. After finishing those steps, we built tables for the tents, stools, just about anything." Finishing this, he took a squad and drained swampy patches of ground. "They made me the company NCO for malaria control."

Other men laid gravel over paved roadways, and in one case built a vehicle bridge over the Piva River.[5]

The Americans had learned a lot about jungle existence since 1942. Company encampments were laid out near well-dug shelters. Some soldiers laid their tents on top of raised platforms because of the hard rains. Chuck Walker, now in command of E Company, provides a fine description of Easy's camp site: "Our first encampment was crude [but] after a week of boredom we were moved to a new bivouac area recently abandoned by a Marine unit. At the tail end of each company street was a small, clear river . . . the current was slow, excellent for bathing." Pete Sherar, one of Walker's lieutenants, was an "adept" scrounger. He managed to trade Japanese souvenirs for a generator, a water pump, and a large washing machine. He got lumber by trading beer for access to one of the saw mills. Each tent had electric lights and a hardwood floor. The water pump fed showers. Colonel Samuel Gee, the battalion's newest commander, ordered the men to plant vegetables from seeds he had obtained in Fiji. Walker planted a tomato patch next to the tent he shared with Sherar.[6]

Every company used small patrols to sniff out Japanese intentions on their front. Leaving on a long-used native track that crossed the Torokina River (the "east-West Trail" as it was marked on maps) larger patrols would fan out for signs of enemy preparations. They found the main trail blocked east of the river by a series of Japanese entrenchments. Beyond that they turned up almost nothing but some empty "spider holes," old ration cans, a few pieces of equipment, and one Japanese corpse. A G Company patrol found signs that enemy troops were somewhere close by. Guadalcanal had demonstrated that the Americans had a decided advantage in firepower over the Japanese, but also had shown that this advantage was diminished when the enemy went to ground.[7]

The green troops still had a lot to learn about survival in the jungle. Few saw any Japanese on these patrols, but their nerves were taut in expectation. Already two men had been shot because they left their foxholes in the dark of night. One was wounded, the other killed, after being "challenged but failed to give the password."[8]

Casualties in these early days were relatively light. Up to March 3, the regiment lost 17 men killed, 98 wounded and 2 men missing. For most men life on the island was fairly safe. George Dingledy, a PFC in Company F, sent his parents a letter describing one of his days in January: "Today we had it a little easy so I took a dip in the ocean. The sun is quite hot so the ocean is quite warm. We have been cut to [mailing] one letter a day for awhile. I guess we are wearing the censor out . . . If you haven't sent the package, please put in a couple pair of khaki camping shorts just like I wore in the scouts. I need a pair of slippers with a good sole, size 9 and ½ . Here's a funny request but put in two sheets, single width, and more white sox 11 and ½. You sweat so much at night a sheet is a good thing."[9]

Dingledy was another of the many replacements. Growing up in Ohio, he enlisted with several of his friends in March of 1943, at age nineteen. "We had discussed options to make the best deal possible when entering the armed forces and laughed that 'none of us wanted to end up digging ditches in the infantry.' We thought getting ourselves on the voluntary list would help. There was one sobering moment when we were sworn in, the officer read us the Articles of War and said 'don't give us any trouble or desert or we'll put you in front of a firing squad' and he looked mean enough to be in earnest." Off to basic training they went. Dingledy and his friends' hopes to stay together were crushed early; most were separated immediately after basic. Only two of his friends managed to join him in volunteering for paratroop training in Georgia, and that lasted for only two weeks before the Army decided they "weren't the material they were looking for."

Soon they and many others were sent off to Fort Leonard Wood, Missouri, to become part of the newly formed 75th Infantry Division. Here Dingledy thought he had found his home for the war. But this, too, came to an end just before the 75th was to be shipped to Europe: "a decision had been made to transfer about four hundred men to army units in the South Pacific [as] needed replacements due to early casualties at Guadalcanal and effects of tropical diseases." Within a short time Dingledy was on his way to the Pacific. He hoped to be assigned to duties in Hawaii or Australia. But that dream was also squashed when officers aboard their transport summoned them above decks to look at "two life sized models of Orientals, one was Chinese and one was Japanese . . . and we were to study them and be able to tell the difference." A few days later Dingledy learned that their immediate destination was New Caledonia and then almost certainly a combat unit after that. "Well, it was the fourth day out and they knew that whatever they told us [now], we couldn't swim home."

Dingledy was "only twenty, a scared soldier away from home for the first Christmas ever." He could not help but think of his family back in Youngstown. "What were they doing? Was it snowing? Were they thinking of me? I was certainly thinking of myself and them." Landing in the high surf of Bougainville after dawn, he looked carefully at the "black sand, dense vegetation, and palm trees." He and some of his new comrades asked someone how far it was to Mount Bagana, the active volcano, "spewing dense, white smoke against the background of a deep blue sky," but a marine warned them that "the Japs held the surrounding ground." They settled into the F Company tent camp, "out in the boonies somewhere," where trees were high enough to block the view of the hills and even Mount Bagana and "some screwy looking birds" made irritating noises. In the next few days, he went on a few short patrols, never once seeing an enemy or a native. Then his company captain came around the tents and asked if anyone could type. "I'd already learned not to volunteer for anything, even information, and no one else said anything either. But he got mad and said 'alright, some of you learned it in school, let's see some hands.' I reluctantly put mine up and a couple of others did. He pointed to me, maybe because I was the thinnest guy, and sent me to the personnel tent. The army runs on records and that's what I did, getting morning reports, ration reports, all that."

The days began to drag by; there was plenty of paperwork, only a few enemy air raids, and little to do when he was not at headquarters. He read some GI editions of popular novels, got an occasional issue of *Time* magazine, "printed on onionskin paper about half the size of the regular copy with no advertising," played some volley ball with other men. "If you were lucky enough to get to a tent with a Zenith multi-band radio then you could listen to the news each night from San Francisco." His officers were pretty decent, telling the men that they could write and post their own letters without censorship. "We were on our honor not to reveal anything, and I was too scared to think about doing it."

To avoid the occasional night raider, he and three other men deepened their own air raid shelter, put in some floor boards and started sleeping below ground. "It was all sand-bagged, and we'd play hearts down there, sometimes all night." After two months on the island he wrote home: "There is nothing new here, I just knocked off some chow. Everybody that writes puts me everywhere from New Guinea to Hawaii, and I got another letter the other day from an unsuspecting soul who said he was positive that I am in Italy. But as yet no one has guessed the correct location." This was just the beginning of what Dingledy called his "long journey [that] was to end twenty-five months later." His words reflected what most men felt as to their experience in the Pacific, a long and strange exile from all that they knew, an alien world that they couldn't wait to see the end of.

Melvin Bork worked similar duties at the 3rd Battalion Headquarters. He found the conditions on Bougainville to be far better than those on Guadalcanal. "It was easier to get around, thanks to better trails and roads, and I wasn't as worried about snipers on Bougainville." Bork did his work under a succession of new officers. "After [Colonel Robert] Hall, I kind of lost track of who all the officers were, they didn't stay as long as Hall and earlier ones." Occasionally, he was sent up to an outpost to deliver messages, or posted on guard against possible enemy infiltration, but he never had to use his rifle. Like Dingledy, he spent much time playing cards, reading, swimming, and just plain killing time. "It took an awful long time to get anything from home. We just spent a lot of time wondering how long this was going to go on."[10]

Rudy Edwardson, the now seasoned veteran in F Company, wondered, too. "If we weren't out on patrols we were often just bored." Sergeant Edwardson was prepared to turn a blind eye when his men got into some mischief. At one point, a few men got their hands on some wire cutters and used them to get into a Seabee warehouse, taking a motorcycle, an outboard motor, a generator for putting up some lights, and various other odds and ends. "We used the motorcycle to run up and down a gravel trail out by our tents. The officers got a complaint and came looking for it, but we just hid it in the jungle bush. They never found it. I think we just tossed the motor away after we couldn't find any boat for it." Both Bork and Edwardson recall men taking bets on when Mount Bagana would finally erupt.[11]

The 164th had been overseas for almost two years by this time, and the longest serving veterans felt that it was someone else's turn to hammer the Japanese into surrender. There was talk of the military developing a system for releasing soldiers who had already served a long period overseas, but for now the men simply had to wait and do their duty.

The Japanese on Bougainville were probably just as homesick as the Americans. But they were girding themselves for battle. Lieutenant General Harukichi Hyakutake, having failed to destroy the Americans on Guadalcanal, made a plan for driving them off Bougainville from his 17th Army headquarters on the north shore. Having ordered his soldiers on the spot to harass the Americans, he gathered a suitable force to take the offensive. He intimated to American interrogators after the war that he ordered the attack under pressure from superiors and that he expected the attack would fail. If this was true, then once again Hyakutake callously sent his men to their deaths in a lost cause.[12]

The best unit that Hyakutake had for an offensive was the 6th Division, a veteran unit that had served in the Chinese theater for years and had carried out the barbarous sacking of Nanking in 1937. The 6th's commander, General Masatane Kanda, would command the assault with his forces reinforced with additional battalions. Hyakutake wanted the American airfields at Torokina pounded into inactivity, and so he added to Kanda's force heavy artillery that would have to be hauled through the jungle. In all, Kanda had about 15,000 men under his command. The attack was to begin in late February or early March, as soon as the Japanese could get into position on the hills overlooking the American lines. In a trek as grueling as the march made in October 1942, Kanda's men built makeshift bridges and carried virtually every shell, every bullet, every tin of food, and every vial of medicine on their backs. General Kanda himself made the journey on his own magnificent stallion while the men hauled oats, the general's tent, and his personnel supplies.

For each pound of supplies that Japanese soldiers received in the Pacific islands, American soldiers got two tons. An American military aircraft came off the assembly lines in the States every five minutes by 1944, six times faster than in Japan. Japan's own fighters and bombers in Rabaul, assailed by MacArthur from New Guinea and from Guadalcanal and Bougainville, were already a fraction of what was needed to engage their enemy on anything like equal terms. But the soldiers, raised in strict discipline and indoctrinated by the Army to fight on until dead, were willing to trade their lives for a victory. "To attain our aims we must always attack," Kanda exhorted his men. "Serve in silence and bear all pain. The shame of our souls will give us strength."[13]

By the end of February American commanders knew the enemy attack was imminent. Bits and pieces of information gleaned from intercepts and reconnaissance flights provided some idea of how many enemy soldiers were on the way. Army interpreters interrogated some prisoners brought in from patrols. Interrogators in the 164th S2 section obtained additional information. One prisoner "named the 3 [Japanese] inf. regts to take part in the attack plan" and "gave data on weapons, rations, morale, and personalities" of commanders. Another prisoner said he was positive the attack would begin around the tenth of March.[14]

On the map the American lodgment looked something like the blade of a GI shovel, about 11,000 yards wide at its base along the beach, 8,000 yards deep from the beach to the point of the shovel, and about 23,000 yards along the perimeter line. Twelve battalions of the Americal and 37th Divisions were ranged along the perimeter, each holding a front ranging from 2000 to 2500 yards. The 164th Infantry was positioned on the right of the 37th Division. Because the 145th Regiment of the 37th was occupying Hill 700, the 164th's left flank was part of a small salient that the Japanese might attempt to envelope. This point in the line was reinforced by mines, booby traps, and concertina barbed wire, and the men were kept on alert. The 164th positions extended eastward over the forward slopes of several hills, then followed the west bank of the Torokina down to the beach (see Map 7). Most all of the rifle pits and gun positions were protected with sandbags and earthen log coverings.

General Griswold and his staff felt certain that the Japanese thrust would come from the hills north of his perimeter. Griswold was tempted to launch a series of attacks of his own. But he reasoned that, at the cost of valuable men, he would just extend his own line and thin out his forces, while the Japanese would attack from hills just beyond the ones he had taken. So he placed a composite unit of riflemen with artillery observers on a Hill 260, about 300 yards east of the perimeter, ordered the existing positions on full alert and waited for the blow to fall.

The attack began in the hours before dawn on March 8, when the Japanese guns opened a withering barrage. Rather than target the American's perimeter defenses, Kanda hit the American airfields. He had to keep the American fliers from getting into the air and raining bombs on his own infantry. The Americans were surprised by the force of the Japanese artillery. "We didn't think that they could drag those guns over those jungle trails," Alvin Tollefsrud later remarked. Bill McLaughlin, a gun commander with the Americal Division's artillery, had to admit that he admired the Japanese for "doggedly [hauling heavy guns] over the mountains and through thick jungle growth." With his practiced eye, McLaughlin saw quickly that the enemy guns were concentrating their fire on the airfields. He saw as well that while shell "craters were opened up in the strips" they were "quickly filled in" by Seabees. He directed his own guns on the enemy artillery.

Over in E Company, Chuck Walker scoured the hills outside the perimeter, looking for gun flashes. He and a few men spotted flashes coming from trees over across the Torokina River, and a 37mm anti-tank gun was used to fire at the spot. After several rounds, no further signs were seen of the enemy gun. Soon after this he was called over to the E Company tents. Several Japanese shells had pasted the bivouac area. "Our supply tent was gone, the area was littered with broken rifles and equipment. Our artificer, Joe Lauer, was digging frantically into the mess with a shovel." Lauer told Walker that John Wells, the supply sergeant, was wounded and buried under the rubble. Lauer kept digging and Walker ran to the battalion tents to find a doctor. On his way a heavy Japanese shell landed right beside him, a fortunate dud. At the HQ, Walker had to pull a doctor out of a shelter and push him back to help Wells. "The poor guy was white as a sheet and trembling like a leaf under the bombarding fire." Lauer had extricated Wells whose buttocks had been turned into a "bloody mess" by shrapnel. The doctor treated him and Wells returned to duty after a couple of weeks.[15]

The shells fell off and on throughout the day. One shell hit the officer's mess of the Americal Division's HQ, killing two men and wounding three. But the Japanese infantry did not attack until the enemy finished probing the American lines for weak spots. Patrols along the wire fought brief clashes with small numbers of Japanese, killing several men and directing mortar fire onto others they spotted.[16]

The Japanese infantry attacked after dark. Four battalions of Kanda's infantry came out of the bush and rushed forward in a drenching downpour to charge up Hill 700. They broke through the first section of barbed wire, and dashed on toward the second section. Between the two sections the American had seeded the ground with booby traps, established firing lanes and registered the ground for artillery. Two companies of the 145th Infantry waited on the hill top with forward artillery observers who called for fire from four battalions of howitzers with the 37th Division, two more from the Americal. Thousands of explosive rounds fell into the killing zone. A prisoner later told interrogators that his battalion was "practically annihilated." An American correspondent present described the enemy dead as "a gruesome, bloody, stinking human mat on the barbed wire entanglements." Some Japanese survived by finding defilade positions in the ground or by creeping as close as possible to the American lines. After two hours the barrages lifted. Survivors emerged from cover and resumed their advance, some throwing their bodies across the wire while others used explosives to blast their way through. They charged at the GIs in their foxholes.

Now 60mm mortars responded and poured another rain of fire on the enemy, but they still came on and vicious hand to hand fighting broke out in the darkness. One machine gun crew on Hill 700 fought a losing battle against a mass of attackers, mowing down over a hundred Japanese before the four men were themselves killed. A sergeant in another position had placed buckets filled with oil along his part of the line, which he now ignited with phosphorous grenades. The sudden light confused the enemy long enough for his men to cut down dozens of them. Small desperate battles broke out along the line as men fought with their weapons, grenades, rifle butts, knives, and fists. Reinforcements rushed up to the hilltop and stemmed the Japanese tide, but by dawn the enemy held a small penetration about fifty yards deep and seventy yards wide. They expanded their wedge slightly in the morning hours. But the Americans captured priceless pieces of intelligence from the body of a Japanese officer who had carried with him "complete plans for the attack."[17]

To the right of Hill 700, men in the 164th had spent the night on alert. At one point, some shells fired in support of Hill 700 fell short, close to I company and the 3rd Battalion headquarters position. "The art[illery] was notified and changed range," regiment records noted laconically; one suspects the request was rather more forceful than that. Men in Company K were able to lean out of their foxholes and watch their own artillery duel with the enemy guns. One witness noted that the "Division Artillery cleared the camouflage around [some of] the enemy guns so that the emplacements could be seen" on the ridges beyond the perimeter. Men who had binoculars watched "Japanese frantically trying to repair the camouflage" but the American shells killed them and destroyed a gun. The men remained in their positions throughout the night, but except for "what sounded like chopping and muffled voices" beyond the wire, the 164th lines were very quiet. Then in the morning, Horace Nearhood, a young replacement from Michigan now in B Company, looked out and saw "a Jap that got through our wire and crawled up to our bunker. We all let loose, and I probably fired the whole clip in my BAR, but one of our sergeants shot him. I think maybe he was trying to surrender but didn't know how to do it."[18]

Americal Division lines came under attack on the morning of the 10th, when another battalion of Kanda's division attacked Hill 260, overrunning an artillery observation post and some pillboxes. In order to call down artillery support, Frank Sager, a spotter with Battery C of the 246th Artillery Battalion, took a radio and went forward with Sergeant Fred Squires. "We were to be their forward observers for artillery support against the Japanese infantry trying to scale the slopes of that advance position." Sager set up his radio in a slit trench and placed a grenade alongside to destroy it "if we were killed." As he called down fire against the charging enemy soldiers he watched the battle unfold. "I saw one of our machine gun men get hit badly. A young GI (later I found out in the hospital in New Hebrides that his name was Gordon) jumped right on that machine gun and did a 'John Wayne' spraying lead into the Japs charging up that hill." Despite a stout defense, the Americans on the eastern slopes of the hill had to fall back. Sager withdrew, but suddenly "felt this terrible pain in my back and right hip." A mortar fragment had struck him and he could not stay on his feet. Squires left him briefly to find a medic. Two advancing enemy soldiers passed him thinking he was dead. Squires returned with a medic and Sager was carried back into the perimeter.[19]

The Japanese pushed on and grabbed a key feature on the hilltop, a giant banyan tree some hundred and fifty feet tall. The 182nd used it as an observation post, rigging up a navy bosun's chair for the spotters. From the tree Japanese could see the entire perimeter. General Griswold ordered General Hodge to "hold hill 260 at all costs." Two more companies of the Massachusetts regiment sent to drive the Japanese off the hill managed to gain no more than about twenty five yards before they were stopped cold by machine gun fire. A second attack with flamethrowers also stalled. Both companies took heavy losses with one down to about half its normal size.

Brigadier General William McCulloch, the assistant division commander, went forward to take direct command of the battle for Hill 260. He wanted the giant banyan tree observation post destroyed by massed artillery fire. He also called for the provisional flamethrower platoons of the 132nd and 164th regiments. Flamethrowers were still very new weapons. Only a few men had been given the training to use it properly, and they all were wary of it. "They were frightening weapons," an operator on Bougainville told a military historian after the war. "When you use a flamethrower, the blast gives off an intense heat that burns off your eyebrows. It creams everything it hits, just destroys it." With large tanks of flammable jelly and propellant on his back, the operator was easy to spot, and he became the bull's eye of intense counter fire.[20]

Two rifle platoons of G Company accompanied the 164th flamethrower team to Hill 260, where they fought in a brisk, three-day fight reducing Japanese strong points. "Heavy casualties were inflicted" on the enemy, the regimental history noted, "but our forces also sustained sizable casualties. Indeed, the regimental flamethrower team lost nearly half their men. The 132nd teams lost almost as many and the 182nd teams were all but wiped out.[21]

It took seven days of battle for the Americans to regain control of the hill. Over ten thousand artillery shells from American guns had plastered the hill and still enemy machine gun nests had to be eliminated with explosives, rifle fire, and flame. As casualties mounted, Donald Blaisdell, a private with the 57th Combat Engineers, hauled wounded men back to the aid stations until he was killed. A Japanese-American (Nisei) interpreter, Shigeru Yamashita, put on a Japanese uniform and spied out the enemy positions. Progress was slow but steady. The brunt of the fighting was borne by the 182nd and Company G of the 164th. Toward the end, American engineer troops used pipes to pump gasoline into suspected nests and ignite them with grenades. The Japanese left over 600 dead on the hill when they finally withdrew. The American losses were 98 dead, two dozen missing and about 600 wounded. The 182nd's Company E ended the battle with only 24 effective men. Company G of the 164th fared better but still had losses. Combat engineers brought down the giant banyan after men took photographs of the "million dollar tree."[22]

During this bloody mêlée, the 164th waited for its turn to be attacked. Crump Garvin expected that surely the enemy would strike against his middle part of the lines. But no attack came. Instead, Kanda launched another attack against the 37th Division. This started on the night of March 12 with a full-fledged banzai charge at the American positions on "Cannon Ridge" just west Hill 700. Searchlights provided by a coast unit bounced their beams off low clouds and lit up the Japanese while "the concentrated automatic fire of every infantry weapon in the book mowed them down." The battle continued sporadically for several more days. Good men died in "vicious fights" but the Japanese were failing to gain a breakthrough.[23]

Men in the 164th's lines waited and dodged snipers and shellfire. Tony Czarnecki, a young rifleman in Company K, spotted one gun that the Japanese would wrestle out of a cave, fire two or three times, then pull it back into cover. Marine Corsairs tried to hit the cave with bombs. Then the soldiers brought up a 90mm gun mounted on a halftrack and fired on "the puffs of smoke." This knocked the enemy gun out. John Tuff scavenged a 20mm automatic cannon from a downed Japanese fighter, attached a naval gun sight to it, and poured streams of shells into several targets. Some 40mm anti-aircraft guns were directed against suspected positions.[24]

As enemy artillery fire continued, many of the rear area men kept to their shelters as much as possible, some hesitating to come out for food. Alvin Tollefsrud remembered one fellow who decided to rely on canned rations rather than stand in a chow line, saying "I'm damned if I'll stand there and write my name on a shell." George Dingledy and friends played bridge and reminisced about home in their shelter. But then he "picked the damnedest time to get a toothache . . . I hitched a ride over nearly empty roads and somehow found the dentist. His 'office' was a tent in a field near a company street." So as enemy shells dropped haphazardly around the area, the dentist filled his tooth using a drill powered by an assistant on a stationary bicycle. Both Dingledy and the nervous dentist were "all covered with sweat" by the time the tooth was filled. "But it must have been first class [work] since the filling lasted well over thirty years."[25]

General Hodge ordered patrols to take the fight into enemy lines. Chuck Walker took a large part of his company out toward the Torokina. There one of his platoons saw "a group of enemy crossing back and forth on the almost dry riverbed in the rear of Hill 260." Informed of this, Walker brought up more men to attack the column, killing several men and thought that with artillery support he could cut off and destroy a large number of the enemy.[26]

Outside the perimeter, Sergeant Doug Burtell was with the long range recon platoon at the Laruma River, looking for any signs that the Japanese were trying to bring up reinforcements. Burtell and the others found very little to this effect but did pick up an island native. They had been briefed about the natives' early support of the enemy and had heard that natives had seized downed American and Australian fliers and turned them over to the Japanese; they couldn't be trusted. So they kept the native with the patrol until he escaped one night when the man on watch dozed off. Crump Garvin was angry when Burtell reported to him. "He was upset because we hadn't killed the native. I told him we were not about to shoot an unarmed native." Garvin let the matter go, but the incident bothered Burtell thereafter.[27]

By the last week of March, General Kanda had lost heart for the success of his attack. His thrusts at the perimeter had been checked at every turn. The penetrations achieved on Hills 260 and 700 had been pinched off and destroyed. He ordered one of his colonels to take the last remaining reserves for one final charge at the positions of the 37th Division. Some 3000 still game Japanese soldiers tried to break through at a defile on the ridge west of Hill 700. They again were shot down in droves, but did get a small toehold on the ridge. The 37th counterattacked with veteran riflemen, reinforced with "cooks, drivers, and mechanics," anyone who could carry a weapon. Supported by tanks, they drove the Japanese back in brutal fighting. A correspondent for *Yank* wrote that "the infantry [was] as usual footing the bill" for victory. Now totally spent, low on ammunition and almost completely out of rations, Kanda reluctantly ordered a retreat.[28]

A day or two later, patrols from the 164th's K Company went forward to inspect the artillery that Kanda had to abandon. "About thirty guns were counted with an unknown number buried or removed from the area. . . . Hundreds of tin cans indicated that meals had been eaten here and that consumers had hastily departed." Tony Czarnecki, lugging a BAR and a dozen clips, remembered sweating enough to "starch my dyed t-shirt almost white, and you really couldn't quench your thirst because the tablets for the water gave it a sulfur taste."[29]

The enemy fell back, and an uneasy peace settled onto the perimeter. The end of the war in the Pacific was nowhere in sight.

(Endnotes)

[1] Richard Stevens, "A Christmas to Remember," *164th Infantry News*, December 1997.

[2] S2 Journal entries for January 1, and 6-12, 1944, and "History [of 164th Infantry], 1 January – 3 March, 1944," pp. 193-93; both 164th Infantry Association Records; Gailey, *Bougainville,* p. 134.

[3] The Navy beach operations are described in Morison, *Breaking the Bismarck's Barrier*, pp. 360-65. Gailey, pp. 138-39, and Robert C. Muehrcke, *Orchids in the Mud,* pp. 218-28, for the Hornet's Nest battle.

[4] John Miller, *Cartwheel: The Reduction of Rabaul* (Washington DC: U.S. Dept. of the Army, Office of Military History, 1959), pp. 265-269; Willard Garris to Frances, Jeanne and Michael Lanning, February 10 and 28, 1944, Francis and Jeanne Lanning Papers, East Carolina Manuscript Collection, J. Y. Joiner Library, East Carolina University.

[5] Joe Castagneto, with Eugene Scolavino, joint interview with Shoptaugh, September 2007.

[6] Walker, *Combat Officer*, pp. 84-86. David Weiss, a Seabee on Bougainville, described how tents were laid on platforms and how GIs gave him and his friends Japanese souvenirs for building such platforms for them – telephone interview, June 29, 2007.

[7] S2 Journal entries for December 25, 26 and 27, 1944, 164th Infantry Association Records.

[8] Ibid.

[9] Casualty figures in "History of the 164th Infantry, 1 January – 3 March, 1944," p. 188, in 164th Infantry Association Records. Here and below, Dingledy, *I Remember* (1992), supplemented by Dingledy interview with Shoptaugh, September 14, 2007.

[10] Bork interview with Shoptaugh.

[11] Edwardson interview with Shoptaugh.

[12] Hyakutake's interview in the United States Strategic Bombing Surveys, Interrogations of Japanese Officers, is a fine example of how high-ranking Japanese officers frequently obscured their actions when questioned after the war. Gailey, pp. 140-43, suggests that Hyakutake may have believed that only one American division was in the Torokina perimeter, which if true speaks poorly for his reconnaissance.

[13] Gailey, p. 145, and Morison, *Breaking the Bismarcks Barrier,* p. 352; Max Hastings, *Retribution: The Battle for Japan* (New York: Knopf, 2008), p. 58, for comparison of U.S. and Japanese industrial output and logistics.

[14] Report of the Operations of the Division Language Section, Summary of PW Interrogation Reports Between 28 Dec. 43 and 28 Apr. 44 Inclusive, pp. 1-3, 164th Infantry Association Records.

[15] Tollefsrud interview with Shoptaugh; McLaughlin, *Americal Generation,* pp. 95; Walker, pp. 90-92.

[16] S2 journal entries for March 8, 1944; "For an Ex-CHS Track, An Award for Heroism," [1944], undated news clipping from the Concord [Massachusetts] High School newspaper, in the Frank Doe Papers. Doe was awarded a Bronze Star for aiding men wounded in the artillery hit on the officer's mess.

[17] Stanley Frankel, *The 37th Infantry Division in World War II* (Washington DC: Infantry Journal Press, 1948), pp. 144-147; "Dead Japs Form Human Mat on Barbed Wire Defenses," 1944 story reprinted in *164th Infantry* News, December 2000; S2 journal entry for March 8-9, 1944.

[18] "The Great Bougainville Artillery Battle," *164th Infantry News,* August and November 1993; Horace Nearhood, interview with James Fenelon, September 22, 2000 (copy courtesy of Mr. Fenelon).

[19] McLaughlin, *Americal Generation,* pp. 95-96. For an account of artillery fire involving Company K observers of the 164th, see "The Great Bougainville Artillery Duel," August and November 1993 issues of *164th Infantry News.*

[20] "History of the 164th Infantry," entry for 14 March, 1944. The description of the flamethrower in action is in Eric Bergurud, *Touched with Fire,* p. 372. Among the men interviewed for this work, only Bernard Wagner noted that he had commanded a platoon with a flamethrower team which was used to attack enemy bunkers on Bougainville. He declined to say much about the experience, except to note that it was "very unpleasant."

[21] "History of the 164th Infantry," entries for 15-18 March, 1944.

[22] Gailey, pp. 156-61; Miller, pp. 366-72; *Citizens as Soldiers,* pp. 300-302; Col. William Mahoney, "Operations Report on Hill 260," reproduced in the *164th Infantry News,* October 2008; Bruce Jacobs, "The Jungle Fighters," *164th Infantry News,* June 1995. The battle for Hill 260 remains a matter of controversy. Someone in the Americal Division headquarters told historian John Miller that Griswold's order to retake the hill "came as a surprise." They had expected to withdraw to the perimeter once the Japanese attacked, and doubted that the objective was worth the cost to hold it.

[23] "History of the 164th Infantry," entry for 14 March 1944; Frankel, pp. 155-58.

[24] Tony Czarnecki, telephone interview with Shoptaugh, August 22, 2007; Tuff information from Shoptaugh interview with Charles Walker, January 28, 2009; "The Great Bougainville Artillery Duel," *164th Infantry News,* November 1993.

[25] Interviews with Joe Castagneto and Alvin Tollefsrud; Dingledy interview with Shoptaugh, and Dingledy; "I Remember," p. 19.

[26] Walker, pp 93-95.

[27] Burtell, "Patrolling on Bougainville," *164th Infantry News*, June 2000, with further detail from Burtell interview with Shoptaugh.

[28] Frankel, pp. 163-65; Cooper and Smith, *Citizens as Soldiers,* pp. 305-6; Ralph Boyce, "The Screwy War on Bougainville," reprinted from *Yank* in the *164th Infantry News*, June 1996.

[29] "The Great Bougainville Artillery Battle (Second Installment)," *164th Infantry News*, November 1993; Czarnecki telephone interview with Shoptaugh.

Chapter 12: Victory by Attrition
Bougainville: April-December, 1944

With the Japanese falling back, General Griswold ordered both the 37th and Americal Divisions to pursue, "destroying the maximum number of [the] enemy and his material." General Hodge, commander of the Americal, ordered G and F and E Companies to move out, secure some of the hills north of the perimeter, capture or destroy any artillery left there, and cut off the retreating enemy. The patrol advanced across the Torokina, with G Company following the river's east fork while F Company moved on toward the hills. Rain reduced trails to globs of mud. In order to climb the hills, men grasped at vines and crawled upward by inches. "Christ, what a rough go it was at night," Burtell wrote in a short diary he kept that month.[1]

F Company moved up toward Hill 250, where the regiment's 1st Battalion had fought a heated action with the enemy in early January. Now the Japanese were waiting for them again. E Company was advancing toward the crest of Hill 600 in an effort to get into a position where they hoped to trap the Japanese. Chuck Walker got his men onto Hill 600, which gave him "an excellent view of Mount Bagana" and "the east and west forks of the Torokina River." It was about mid-afternoon when his observers spotted Japanese infantry beginning to cross the river. Artillery liaisons went to work and called down fire on the crossing sites.

Doug Burtell, who was along with part of the regimental recon platoon, watched the fire come in on the scurrying Japanese. "What a sight to see that artillery right on them," he wrote in a brief diary he kept at the time. Walker, observing the fire of the American guns, was shocked to see one enemy officer cross the river on a large white horse and identified him as General Kanda. Burtell thought he was foolishly arrogant, but Kanda got across without a scratch. Soon after, a Japanese prisoner was brought in who offered to show Walker a route up to Hill 1111.[2]

F Company in the meantime had come upon the enemy soldiers entrenched on Hill 250. The enemy group was estimated at "200 Nips . . . digging in," left to hold up the Americans. Here a short and vicious fight began. Sergeant Miles Shelley and his platoon with F Company went forward. Just days before, Shelley and his platoon carried out a joint reconnaissance with the platoon of John Paulson, his cousin. They had run into a Japanese ambush in which two men were killed and Miles was slightly wounded. "My cousin [Miles] was hit twice, light wounds in the neck and shoulder," Paulson recalled. "I was nearly hit too but "Big Ray" Anderson shot the sniper first. We got another wounded man out and back to the hospital, but he died. Miles took it hard. He shouldn't have gone out on that next patrol. But he didn't want anyone else to be the lead scout after that because he had the most experience."

Shelley decided to move closer with one other man and look over the enemy positions, but a well concealed Japanese machine gunner opened fire on them. Shelley was seriously wounded while the rifleman fell back and took cover. Shelley called out that he could not crawl back. Men from Shelley's platoon came forward and attempted to reach him with BAR cover fire. But they were driven back by the machine gun. Shelley was wounded a second time by a Japanese grenade blast, but continued to call out to his men, giving directions to enemy positions that he could see. Suddenly a Japanese officer came out of the brush with a group of men and led a head-on attack. The F Company men brought the charge to an abrupt halt, killing the officer and about a dozen others. The rest then fled, some toward the river where men of G Company killed several.[3]

Back in F Company, medics went forward to help Shelley. But he had died. Enemy snipers were still active and this made it difficult to get the dead and wounded back to the company headquarters area. John Paulson had been engaged with his own platoon, and only later learned about Shelley's death. "That was really hard. He wasn't supposed to be there, he had a sore shoulder, but he didn't want anyone else to take the risk."[4]

Vehicles could not reach these hills so the wounded men would have to be hauled back on litters. Men from F and E companies went to work with shovels, bayonets, and helmets, cutting steps into the slick hillsides. "These hills were really rugged," 'Sandy' Sanderson recalled. The medic stayed close to two of the four wounded men who were the most seriously hurt. "We used shelter halves to carry them. One had grenade fragments in a belly wound, he needed surgery. We had to line much of the company up along the trail, over and down the hills, and we passed the stretchers hand over hand, one to another all night, until about 3 o'clock in the morning." Flares dropped by an American dive bomber helped the men see as they continued passing the men down the hills in a perpetual bucket brigade. After almost thirteen hours, they reached a spot where a jeep could carry the wounded men the rest of the way. Cooks brought up hot coffee and cold pancakes. At a reunion many years later, Sanderson learned that all four of the wounded men had survived.[5]

Company E meanwhile pushed on to Hill 1111, where a scout was killed when he made the mistake of moving down the center of a trail. The trail blocked, Walker withdrew his men for the night. It was a cold night, everyone spending it "in a soaking wet blankets" and getting little sleep. Come dawn Walker sent two platoons across a creek to move around and flank the Japanese. Lieutenant Charlie Ross led one of these platoons, which attacked an enemy machine gun. The Japanese then fled up the trail. But as Walker moved up with the lead scouts and made contact with Ross, they spotted two Japanese over in a stand of tall grass. "Get Sergeant [Lester] Kerbaugh, be quick," Walker called to one man. Kerbaugh came up and began to spray the tall grass with his Thompson while Walker, Ross and some riflemen closed in. Walker saw "a man's hands through the leaves" in one part of the jungle and took aim with his own rifle. But he pulled back when he heard the pop of a grenade being primed. The grenade flew by him and past Kerbaugh, exploding harmlessly. A second followed within seconds and exploded close to "Ross [who] got a fragment in his leg." Searching the jungle, Walker's men found blood strains but no enemy dead. After five more frustrating days chasing after Kanda's retreating force, Walker brought his men back to the lines on April 8.[6]

Regimental headquarters was pleased with the pursuit. Losses, they felt, had been "minor." Men in the ranks felt differently about that. Miles Shelley was posthumously awarded a Distinguished Service Cross, but that was of limited comfort to his friends.[7]

The regiment established an extensive "outpost line of resistance" along the crests of Hill 250, 600, 1000, and 1111, and manned this by rotating companies throughout the spring and summer months of 1944. Patrols would go out to gather intelligence and break up any enemy concentrations with ambushes. Aggressive patrolling whittled down the enemy forces and disrupted his ability to recover. But it was dangerous, frustrating work, the more so because it seemed to many that the Japanese had lost the fight on Bougainville but again would not quit.

Emil Blomstrann was one of many replacements who come to Bougainville with Company E. Trained as a rifleman, he was placed with a machine gun crew and spent most of his patrols carrying extra ammunition. "I fired the gun but almost always at distant things like gun flashes or smoke or suspected movement. I don't know that I ever hit a thing." The only time he remembers any of his gun crew being injured was when one of the loaders burned his hands while changing a barrel. "They offered to put him in for a Purple Heart, but he said no thanks." Emil suffered no wounds or injuries and even avoided malaria. "My only problem is I was a red-head and sunburned really badly. John Wells, one of the dozen or so old guys from Guadalcanal, had his tent near mine. He knew how to get things and scrounged up some t-shirts with long sleeves that I wore for sun protection. We all kidded John after he took that artillery fragment right in the rear end. I have a picture of him 'showing off his wound scar.'"[8]

The Japanese failure in the March offensive had left them crippled. After interrogating the newly taken prisoners and analyzing captured documents, American intelligence estimated that the Japanese were abandoning most of their wounded as they fell back. Generals Hyakutake still hoped that he might receive reinforcements and supplies. But this was a pipedream. Since Guadalcanal the American Navy had vastly improved its capabilities. Bougainville was thoroughly isolated; only a trickle of men and ammunition would slip through the blockade on small barges. Allied flyers were reducing Rabaul to impotence. A Japanese officer interrogated after the war admitted that by 1944 Japanese pilots were being thrown into battle "fresh from [flying] school." Most died after just a few missions. Neither conventional methods nor tactics could provide a reversal of fortunes for the enemy's air power. They would soon turn to the unconventional -- suicide attacks on American ships.[9]

By April, patrols from the 182nd Regiment were harrying the Japanese as they retreated. The 132nd meanwhile were reducing enemy bunkers east of the Torokina. The Solomons campaign was over, at least in the strategic sense. The Americans had their base and Rabaul was no longer a threat. Indeed, as one Army historian later noted, the "destruction of Japanese men, materiel and mobility throughout the Solomons and New Guinea left the Japanese mired in a multi-front war they could not win." But they would not concede defeat. They fought on, contesting every foot of the island. The battle of Bougainville was now a war of patrols, ambushes, and counter-ambushes. It was a contest of small unit actions filled with small gains and great frustrations.[10]

Joining the 164th when he was sixteen, Doug Burtell came of age as a warrior. At the conclusion of the battle at Hill 250, he went with a few other men back to the perimeter, guarding the few prisoners they had taken. "We are mopping up now," he wrote, "Japs leaving everything behind. The last two Jap prisoners we captured died before we could get them in. Too bad, but the bastards deserve it." Burtell took part in destroying some abandoned howitzers and celebrated his birthday while doing it. "My birthday today," he wrote. "20 years old. Will stay until operation is over. The gnats are terrific here." Sixty years later he admitted "I thought we might be at this for maybe another ten years. It was clear that the Japanese wouldn't surrender and we would have to fight all the way to Japan." He felt it would be necessary to kill millions to finish the war.[11]

A lot of men agreed with him. Rudy Edwardson admitted that after Guadalcanal "I'd sooner kill Japs than anything on some days." It the only way he thought would win the war. In the same vein, Donald Jackson, a young naval communications officer, admitted that "hatred [of the Japanese] would be part of my armor" on Bougainville. War in the jungles was brutal, remorseless and unforgiving. It was a war that matured young men very quickly.[12]

Patrol warfare had few rules, but those few were crucial. Men must spread out on patrol, separated by a distance of at least five yards. One hidden enemy machine gun would take out most of them otherwise. Silence was to be observed when the patrol was in proximity to the enemy; hand signals were used to convey most information. Men, especially those at the head of the patrol were told to stay away from the trail center, follow the edge, expose little to a waiting sniper. Men moved carefully, for the Japanese set booby traps and any noise could draw fire. Experienced men ordered the rest not to smoke or even carry tobacco. The smell of a burning cigarette could be literally a dead giveaway. With such guidelines, men could keep mistakes, and casualties, low.

Experienced men learned to prepare carefully for each patrol. This was particularly important for the recon work to gather information, avoid combat, and get back quickly. Long range recon patrols generally avoided canned rations; it was too noisy to open the cans and the jungle heat sometimes spoiled the food anyway. There was at least one occasion when an overheated can burst in a man's backpack, convincing him he had been wounded. Backpacks could snag on the foliage. Burtell liked to take "hard candy, raisins, any fruit and stuff like that and roll it into a shelter half or poncho, tying the ends with some radio wire," and wearing bandolier style. Two canteens of water, a jungle knife and bayonet, and extra ammunition completed the kit. Some officers had men add mosquito nets to their packs, carry tracer bullets or incendiary ammunition, gun oil, entrenching shovels, and small Coleman stoves with fuel. Chuck Walker frequently regretted the paucity of flares. "They could be used to call for emergency fire or to mark a position." Increasingly, men used cloth caps on patrol. A helmet striking a vine or branch could be heard.[13]

Patrols required experience. Alertness was essential and men could not panic at finding themselves well beyond the perimeter. Many veterans did not like to have inexperienced men on patrols. Doug Burtell thought little of a newly arrived medic who accompanied one his patrols: 'Doc' was new man – sure didn't have the old spirit – bitched 'cause he had to go up [Hill] 600 at night. Just an old woman." But when one of the other men was wounded, he also admitted: 'Doc' we brought up helped a lot." Bernie Wagner generally did not like to take any of the "rear echelon" men on his patrols – they talked too much. Men on patrol felt safer, Chuck Walker later wrote, when "each man knew the abilities of the other men in his squad or platoon . . . each man knew his life might depend on the ability of his comrades."[14]

    The lead scout had the riskiest assignment on a patrol. Burtell "was point man plenty of times, too many probably. You had to watch behind every leaf, almost, and it was nerve wracking." Burtell remembered that after returning from a long patrol it was not unusual for him to sleep fourteen or fifteen hours, "and I was younger than a lot of the others." Rudy Edwardson organized supplies for the platoons in F Company and joined patrols as "a floating platoon sergeant." He made certain that all the men assigned to patrols were well armed by building up a stock of extra M1s. "For every kia and wia we had, I reported a lost rifle, which wasn't always the case." When on patrols, Edwardson ordered his men to wear only their light fatigue caps and for rations, he liked to carry "a few chocolate bars and a can of bacon for several men." If 60mm mortars were taken along, he usually carried TNT and primer cord so he could clear a firing zone. "We never carried or made any maps, we travelled by compass and time and those who debriefed us made the maps. On returning they'd give us a shot of brandy and question us after that. Some of the patrols were quiet and more fun, some were like sticking your head in a noose. I felt better if there were a Fiji scout or two with us. They were tough cookies, boy."[15]

Every man who had contact with the Fijians spoke highly of these tough and capable warriors. Fijians could march twice as fast and twice as long as the average American soldier. A Fijian could go without food for longer periods of time. His sense of the jungle was unparalleled. Fijians showed Doug Burtell how hungry Japanese would find food from a small palm tree. "There was a white piece, like cabbage, near the top. If you spotted that, you knew they were damn well around you." Fijians had excellent senses of smell. When one patrol made up of men of the 164th stopped to bed down one night, the Fijian scout calmly pulled out his pipe and sat down for a smoke. Shouldn't he put that out, one of the men asked. No, "if there was one within a mile of here, I'd know it." The Fijians were always in good humor, loved to sing, and seemed to take the war as a great game in which they could prove their courage. They frequently killed sleeping Japanese soldiers by slitting their throats in the dead of night. Some Fijians took body parts as trophies from their kills, although the Americans discouraged this out of concern for retaliation.

Lieutenant Charlie Ross went on several patrols with Fijians. In June, he accompanied an E Company "reconnaissance in force" from the Hill 250-1111 outpost line toward the east-west trail. They ran into a well camouflaged Japanese position of "strongly fortified bunkers" and a serious firefight quickly developed. One hidden enemy machine gun opened up on the patrol's flank and several men were wounded. Corporal Sefanaia

Sukanaivalu, one of the Fijian scouts, crawled forward twice to pull wounded men back toward the medic. But on his third trip he was hit and seriously wounded. His own comrades tried several times to rescue him but were driven back by fire. Twice he told them to leave him but they refused. Unwilling to let them be killed, Sukanaivalu stood up and charged the gun. He was killed immediately. The American patrol was forced to withdraw. The British Army later award Sukanaivalu the Victoria Cross, for "deliberately sacrifice[ing] his own life knowing that in no other way could his men be induced to retire from a situation in which they must have been annihilated." Fijians on a later reconnaissance patrol led by Ross took a measure of revenge when they found a force of Japanese living in grass huts well off one of the trails. They foolishly neglected to post guards. Calling in some help from the anti-tank company, the combined group hit the huts with mortars, machine guns and rifle fire, "killing 26 Nips." The only American casualty was another lieutenant "slightly wounded in [the] leg."[16]

Fijians had great success patrolling at night but the GIs usually found it too difficult to coordinate the men and keep noise to a minimum level after the sun went down. "We have learned that an attack at night does not have a considerable amount of surprise in it for enemy forces," an S3 report noted, and recommended only "select officers and men that are well-seasoned combat troops" could be effective in a night operation.[17]

Medics generally accompanied all patrols. They ran sizeable risks in the jungle, partly because the Japanese did not respect the non-combat status of the medic, but also because the medic could never keep safely under cover. PFC Richard Reid was a member of the 121st Medical Battalion and served with companies from the 182nd Regiment. He vividly remembered his Bougainville patrols out beyond the lines. Reid was a very religious man, inclined to pacifism. He refused to carry a weapon, even after the officers of the 121st urged all the medics to qualify for small arms fire. He recalled that one of the other medics was annoyed at him for this. "He told me that if my refusal to shoot ever endangered him, he would shoot me himself." Reid treated men under fire a couple of times, using morphine and sulfa powder on minor shrapnel wounds, and on one occasion was part of a group that carried some wounded back to the perimeter. "There were about 24 of us used to carry one man, in relays, back about thirteen miles, over creeks and ridges." Ironically he only came under fire himself when back in the perimeter. "One time, a Japanese soldier had infiltrated us and got into some trees surrounding out tents. He tried to snipe us. One fellow was nearly hit as he was entering his dugout and a few bullets hit close to us. Some other guys with rifles drove the sniper off." He administered plasma to several wounded men at the base hospital but never used it in the field, "which was a good thing because it was hard to carry."[18]

Gene Scolavino was assigned as medic to Company A, almost always with the 2nd Platoon. During patrols, he tried to be place himself in the middle squad line of march. "I thought that way I would be about in the center, able to go where I was needed when trouble started." He treated men for a variety of wounds. Most wounds he treated quickly and got the men back to an aid station. But one time, the squad's BAR man, a friend of Scolavino's, accidentally kicked the stock of his weapon while marching. "He had left the safety off, and the BAR went off, tearing up his left side. He died despite everything we tried to do." On another patrol, the platoon sergeant was shot in the neck during a fire fight. He too died. "When men were seriously hit, they almost always asked for their mother at some point."

Scolavino did administer plasma in the field. "The plasma was in two bottles, one of the powdered plasma, the other of purified water. The bottles could break if you weren't careful. You also had to be sure to puncture the bottle of water first, because the plasma was in a vacuum. When you pushed the tube into the plasma, the vacuum drew the water in to mix it. If you did it wrong they wouldn't mix. I saw that happen a few times." Scolavino carried an M1 carbine, but seldom fired it, "just a couple of times in the direction of enemy fire."[19]

A medic could become a patient in seconds. John Anderson was wounded back in the perimeter during a sudden enemy air raid. Japanese planes came out of the rising sun one morning and dropped several sticks of bombs before speeding away at low altitude. Anderson was hit by bomb fragments in his neck, leg, and hands. As he fell, his friend Red Murray jumped on top of him to protect him from further explosions. John's most severe injury was a deep gouge in his left thigh. The whole time the bombing lasted, Anderson was praying "to the Blessed Mother to save him." The attack was over quickly, but a couple men had died and Anderson was seriously wounded. He was rushed to the nearest field hospital where a surgeon then went to work on the damaged leg.

Before he went under the anesthetic, young Anderson heard one of the assisting medics worry aloud that they might "have to amputate." But the surgeon said "no, let's try to save this boy's leg." The next day, he awoke with his leg still attached, heavily bandaged. An air transport carried him to Guadalcanal a few hours later.[20]

Dennis Ferk had become the supply sergeant for Company A, a job that was certainly less dangerous than his rifleman's duties on Guadalcanal. But he too was injured on a hit and run bombing raid. "Japanese planes would fly over the island just after dark to drop bombs. On 10 May '44 we could see that the plane would be close to our unit area." So Ferk and his tent mate Chris Kolden headed for their shelter. "Chris jumped in first and I was right after him, feet first. As I dropped into the bunker, I saw the flash of the bomb, felt the concussion, cleared the logs going down, then heard the shrapnel hit the logs." After the all clear, the two men found their tent shredded by bomb splinters. Ferk's face had been sprayed with dirt and gravel as he jumped in the shelter, but he thought nothing of it at first. Then as he worked on forms at the company supply tent, he noticed he was having difficulty seeing. "I was tired and put my hand up to rub my right eye. I couldn't see a darn thing. I was really shook up."

A medic examined him and quickly sent him over to the 121st Medical Hospital. The doctor at first thought he had a concussion from the bomb attack. They decided to send him to Guadalcanal. Ferk "took some bottles of liquor with me and a few articles of clothing." The general hospital on the 'Canal could determine if his retina had been damaged.[21]

The medical facilities around Guadalcanal included the 500 bed facility of the 137th Station Hospital. It was staffed by some 300 doctors, orderlies, nurses and other staff. The medical service was now so extensive that the natives on the island, still acting as laborers for the Americans, received care at the base. When Dennis Ferk landed at Henderson Field he thought it was ironic that the female nurses on the island were now quartered in huts "very close" to where he had hugged the ground during the battleship bombardment. Ferk entered the hospital straightaway and when the doctors examined his eye, they found a tumor on his left eyeball. "They were afraid the tumor might spread and cause me to become blind in both eyes." The doctors decided to send Ferk back to the U.S. He flew to San Francisco on the series of PBY flights and spent six months having the tumor removed and his eyes treated. He served out the war as part of a "bazooka demonstration team" performing for savings bond rallies.[22]

John Anderson's stay on Guadalcanal was more extensive. Surgeons operated again on his leg and transfusions of whole blood helped to save the limb. He was sent home by ship, his leg protected by a heavy cast that practically immobilized him. Pain from his wound would bother him for the rest of his life. Still, he later told his wife, he "never forgot the joy as they sailed under the Golden Gate Bridge." After more surgeries and rehabilitation, Anderson was given a medical discharge.[23]

The Guadalcanal hospital treated numerous cases of gas gangrene, a fast acting infection that developed when wounds were not properly disinfected. Bill Buckingham, an orderly, helped to treat several gangrene cases with hydrogen peroxide and penicillin. First mass produced in 1942, penicillin saved countless lives in the war. So too did stocks of whole blood for transfusions. Whole blood, stored at the Station Hospital, saved many men who would have died otherwise.[24]

Eugene Solomon, a lab technician who served at the hospital, coaxed men and officers to donate blood. "Blood donors were paid $25. One day I went to the supply depot to recruit donors. One of the enlisted men agreed to be a donor, but when I asked the supply officer, he told me he would not give his blood to anyone. I suggested to him that he'd better hope that he never needed any blood." Solomon spent a lot of his time doing lab cultures for malaria, and instructed men sent out to other islands how to reduce malaria by controlling the mosquito population. "I'd show men how to spray oil on standing water, using a hand-pump sprayer, how to cut down grass with machetes. I also said 'keep the men fully clothed from dusk to dawn. That wasn't easy since it was so hot over there." Serious recurrences of malaria were treated at Guadalcanal.[25]

Men in the Pacific got medical aid for everything from wounds to fever, skin ulcers to concussions, hysteria to venereal diseases. In the year since General Patch had complained that every case of fatigue should be considered "cowardice," attitudes had mellowed. The 137th Hospital had a well organized neuropsychiatric unit. The hospital commander repeatedly asked for more trained Army psychiatrists. The 1944 Annual report for the hospital stated that most cases treated by the neuropsychiatric unit could be intimately tied to "poor morale" which in turn was most often linked to "inequalities between officers and enlisted men" in matters of promotion; liquor, leave, and other comforts; poor leadership; and above all "long service overseas." The report concluded that morale was most bleak among men who had been "overseas for two years or more."[26]

The Rotation Policy that the medical report referred to was indeed a matter of considerable discontent. In February of 1944, the Army had issued a circular stating that "those with the longest service in command or those with the longest service overseas" should be considered most eligible for furloughs. The circular noted that theater commanders had full discretion to make the choices and should in fact consider the "exigencies of the service and [the] prosecution of the war" when releasing men. That clause gave "the brass" the power to hold on to experienced company combat officers, sergeants and corporals. Men who inquired about a furlough were told "sorry, we can't spare you." What it came down to it, only a limited number of men in combat units of the Americal Division got home in 1944.[27]

The Americal's official history admits, "morale waxed and waned among the older men" as a few were granted furloughs, but most still waited, hoping "that next month would be their month." In the meantime, men back in the perimeter relieved their boredom through baseball, volleyball, arranged boxing matches, and glee club competitions. They went fishing, sometimes with poles, sometimes nets that local natives made, and sometimes with explosives. They played cards, shot dice, gambled on just about anything conceivable, including when the island's next earthquake would occur. On one occasion, a large group of men started a pool on who "can grow the most luxurious beard."

Having studied some art before the war, Emil Blomstrann carved items out of exotic woods. He started working on the bust of a woman, but having no model to work with he could not decide how to make it look. "Carrol Cooledge from F Company gave me some movie magazines." From these, Blomstrann chose the actress Gloria DeHaven. John Wells was gutsy enough to write to DeHaven asking for more photos, while an officer somehow arranged to get five pounds of modeling clay for Emil. DeHaven sent some publicity shots and Emil had the bust completed in a few weeks. Men came from across the perimeter to "take a look at it; we were pretty far from home and missed seeing a sweet American girl." Another man who took up carving was Corporal William Freeman, a company clerk who used native mahogany to make jewelry that he mailed to his wife back in Massachusetts. Later, in the Philippines, Freeman won a Bronze Star for helping to rescue a wounded man under fire.[28]

Drinking and griping were common pastimes, one often influencing the other. Men sometimes got a beer ration but traded for more with the Seabees. They mixed up their own concoctions with raisins, canned fruit, diluted torpedo alcohol, whatever. Doug Burtell remembered that Eli Dobervich, a part of the headquarters staff, had promised his fiancé that he would avoid alcohol during the war. "So he'd just sit around and moon over his girl, saying 'I want to see 'Junie-bug.' I got sick of hearing about his 'Junie-bug' and wished he would drink." Drinking sessions usually passed peacefully, but there were some embarrassing moments, including one late night party arranged for officers and nurses. The dance was briefly interrupted when a nurse screamed from one of the latrines. A "drunken officer" had wandered in and "stood at the open door of a toilet urinating on an Army nurse."[29]

During the Japanese offensive, tempers had flared between the headquarters staff and field commanders, and as the months dragged on, ill will simmered on a number of issues. Captain Walker was constantly irked by the special privileges dished out to the "bootlickers" at headquarters. He freely aired his opinion that "the most capable and humane officers were from civilian status or had come up through the ranks," and admitted that by saying this "I buried myself a little deeper" in the eyes of career officers. But other officers shared his feelings. Milton Shedd, a young lieutenant who showed much skill in combat, wrote home in disgust that he had been called to the regimental HQ "for information concerning what my battalion commander did [while on a patrol] that might help him get a medal. Actually he did no more than most of the privates in our company and less than any man on my patrol. I told him I had no information, very politely and referred him to other officers . . . . The number of medals doled out to the brass is out of proportion to that given to enlisted men. If I were a private I would go mad!"[30]

The war had greatly changed the regiment from what it had been in 1941. In M Company about seventy percent of the men by mid-1944 were transfers and replacements. Most of those who had been PFCs or privates at Camp Claiborne and were still in the 164th on

Bougainville were now either officers or sergeants. Ten of the fourteen staff sergeants had been privates or PFC at Claiborne. This type of change was discernable in all the companies.[31]

The old nettle of "the Guard against the regular Army" still existed but the changeover in personnel had relegated it far behind a man's worth in combat. If an officer was careful with his men's lives, he was respected. If not, or if he was uncaring, arrogant, or too ostentatious with his privileges, then he was generally regarded with contempt. Bernie Wagner's old comrades accepted his promotion to lieutenant because they had fought alongside him when he was a sergeant. They were sad (and a little envious) to see him sent home after he was injured in a jeep accident in mid-1944. Rudy Edwardson, was promoted to company sergeant in Fox Company, and most men like the choice. Edwardson in turn judged his company captain on the grounds of how well his orders preserved lives. "Sometimes he'd tell me what he wanted to do and I'd say 'so the objective here is to get us all killed?' Then I'd make a different suggestion." One officer told interviewers after the war that he relied on the veteran noncoms; "I would have been a damn fool to tell them what to do. They knew more about it than I did."[32]

Experience was highly prized. The Army was developing a "rotation policy" that would permit some of the long-serving enlisted men to get furlough's back to the States. But the plan was not ready yet. Nor did it apply to officers. Al Wiest remembered that commanders "had the notion of not releasing someone they needed badly." He got a ticket home because "they drew some names out of a hat. Tony Beer and I were the first two names drawn. Crump Garvin was mad as hell to lose such experienced guys. He offered us promotions as majors if we would stay but we quickly, firmly and politely said no."[33]

Vince Clauson, having recovering from his Guadalcanal wounds, including the face wound that left him with permanent scars, had returned to his machine gun platoon in Company D. He took part in several patrols on Bougainville, where he admitted that he and his friends "didn't stick our necks out." One or two of the officers complained that the men were not "aggressive enough." "Every once in a while we would lose a soldier, which was very sad," all the more so after the Japanese offensive had failed. Back in camp he and his friends played "endless rounds of pinochle" and talked about home. Twice he and a pal snuck off to the rear area to swim and hang around a PX to play ping pong. Caught the second time, they went before their platoon lieutenant who laconically asked 'who won' and then gave them extra patrols as punishment. Grousing about this, they half thought of "hooking a ride back to the States."[34]

Even for the combat-hardened men, there came a "there but for the grace of God" moment. John Paulson was part of a patrol that found an American fighter that had crashed "in a deep narrow ravine." He thought the plane was a Navy Corsair. The skeletal remains of the pilot were still in the cockpit. "We couldn't see any indications of bullet marks" on the plane or in the cockpit. "Maybe the pilot had been wounded [or] ran out of gas and was trying to make a landing . . . the jungle was so thick in the surrounding area that it is unlikely that the plane would have been spotted from the air." Paulson examined the pilot's dog tags. "His name as I recall on his identification tags was 'Roy Davis.' I put all the information I could on my report." He never knew if the remains of the pilot were recovered. "Not knowing if the dead fighter pilot was ever found has bothered me for 56 years."[35]

In late July, the regiment sent a company up the valley of the Laruma River to relieve elements of the 37th Infantry Division, which had been manning posts far to the east of the Torokina. For the most part, the men at these outposts sent their days probing for any weaknesses or signs of another enemy buildup. The two sides exchanged mortar and artillery fire, and there was the occasional fire fight. Casualties were light, and most patrols returned to their base camp "without incident." But someone's luck would simply run out. Pete Baldino was in C Company, one of the men who had been transferred into the 164th Infantry back in California in 1942. He had survived all the battles at Guadalcanal without a scratch. On Bougainville he had done his share of patrols, again without an injury. The diary he kept was mostly a long list of paydays, letters sent and received, and a few remarks about bombings and shellings. A fairly devout Catholic, he attended Mass whenever he could.

In August Baldino's platoon was ordered to take its turn at the Laruma outposts. Before leaving the main perimeter, he noted that he had the chance to catch a movie, "See Here Private Hargrove," at one of the makeshift 'theaters." Then he bought some Christmas presents for his parents and siblings. Trucked up to the Laruma with his platoon, he filed into a foxhole facing the other end of the valley, a spot where, the regiment's operations journal noted, "small skirmishes took place practically every day." On the 29th about fifteen Japanese launched a brief raid on the American lines. Pete Baldino was killed by a grenade blast. His company commander sent his diary home to his parents. The last entry in it read simply, "wrote to Julie," a girl from back home.[36]

(Endnotes)

[1] S2 Journal entries for April 1, 1944, 164th Infantry Association Records; Gailey, *Bougainville,* pp. 169-70. Gailey stresses that these March-April actions were undertaken to push the Japanese back and further reduce his forces. The contention in *Citizens as Soldiers* (p. 302) that Griswold wanted to "eliminate all Japanese resistance on Bougainville," is overstated. General Hodge was promoted to commander of the XXIV Corps and handed command of the Americal Division over to General Robert McClure while this patrol action was underway. Hodge incidentally told staff that the 164th was the "best regiment of the Americal" – Cooper and Smith, *Citizens as Soldiers,* p. 329. Douglas Burtell statement in manuscript diary, entry for March 29, 1944.

[2] Burtell diary, March 29-April 1, 1944; Walker, *Combat Officer,* pp. 96-97.

[3] Cooper and Smith, p. 303.

[4] John Paulson interview for the North Dakota Historical Society, July 18, 2007, and interview with Shoptaugh, October 20, 2007.

[5] Gerald Sanderson interview for the North Dakota Historical Society, November 21, 2005, and subsequent interview with Shoptaugh, September 2007; Shoptaugh conversation with Charles Walker January 28, 2009; Cooper and Smith, p. 304.

[6] Walker, pp. 98-101. Cronin, *Under the Southern Cross*, pp. 169-172, summarizes these engagements on Hills 250, 600, and 1111 together with related actions by the 132nd and 182nd regiments.

[7] "Remembering Miles Shelley, A 164th Hero," *164th Infantry News,* July 2007. The "History of the 164th Infantry, 1 April – 30 June, 1944," in the 164th Infantry Association Records, lists some 150 total casualties by mid-April.

[8] Emil Blomstrann, telephone interview with Shoptaugh, August 13, 2007.

[9] Eric Bergerud, *Fire in the Sky: The Air War in the South Pacific* (Boulder CO: Westview Press, 2000), pp. 328-29.

[10] Cronin, *Under the Southern Cross,* pp. 174-76, for details of 182nd and 132nd patrols; William L. McGee, *The Solomons Campaign, 1942-1942, Vol. II: From Guadalcanal to Bougainville* (Santa Barbara, CA: BMC Publications, 2002) pp. 476-542, for the U.S. Navy advances in beach, naval and air support. Stephen J. Lofgren, *Northern Solomons* (Washington: U.S. Dept. of the Army, 1993) p. 35.

[11] Douglas Burtell manuscript diary, entries for April 2, April 10, and April 16, 1944 (a transcript of the diary was also published in the December 1999 issue of the *164th Infantry News*); Burtell interview with Shoptaugh.

[12] Edwardson interview with Shoptaugh, October 16, 2007; Donald Jackson, *Torokina: A Wartime Memoir, 1941-1945* (Ames: Iowa State University Press, 1989), p. 6. Jackson's book is an excellent memoir of Bougainville.

[13] Burtell interview with Shoptaugh. Backpacks were not always a nuisance Joe Castagneto noted a time when a man in one of his patrols was saved from mortar burst because it "had penetrated a can of bacon" in the man's backpack; "he didn't receive a scratch." See "Saved by the Bacon," *164th Infantry News*, March 2007.

[14] Douglas Burtell, interview with David Taylor, September 2008, and Burtell interview with Shoptaugh; Walker, *Combat Officer*, pp. 125-26; Shoptaugh telephone interview with Bernard Wagner.

[15] Douglas Burtell, interview with David Taylor, and Burtell interview with Shoptaugh; Shoptaugh telephone interview with Bernard Wagner; Shoptaugh interview with Rudolph Edwardson, October 16, 2007.

[16] General Charles Ross, June 14 2000 statement about Sukanaivalu (copy given to Shoptaugh); Walker, p. 114. See also Oliver G. Gillespie *The Pacific: The Official History of New Zealand in the Second World War 1939–1945* (Wellington, NZ: New Zealand Historical Publications Branch, 1952), p. 286. The Victoria Cross was the British version of the U.S. Medal of Honor. See Cooper and Smith, p. 305, and S2 Journal entry for August 3, 1944, for ambush on the enemy grass huts.

[17] "Night Operations in Pacific Ocean Areas, 164th Infantry," February 23, 1945, 164th Infantry Association Records.

[18] Gailey, pp. 120-127 for general medical care; Richard Reid, e-mail to Shirley Olgierson (editor of *164th Infantry News*), February 20, 2007; Reid, telephone interview with Shoptaugh, February 20, 2007. Reid incidentally noted that after the war, in occupied Japan, the man who had threatened to kill him for refusing to carry a weapon apologized, saying he was "proud to serve with me."

[19] Eugene Scolavino, telephone interview with Shoptaugh, April 12, 2007; Scolavino-Castagneto interview, September 2007.

[20] Frances Sinclair-Anderson to Shoptaugh, July 2007, concerning her husband's injuries on Bougainville; S2 Journal entries for January 23-24, 1944.

[21] "From the South Pacific to 'Here's Your Infantry': Excerpts of the Story of SSGT Dennis R. Ferk, Co A," *164th Infantry News*, July 2007; SHSND Interview with Ferk, September 25, 2007; Shoptaugh interview with Ferk, September, 2008.

22 "Annual Medical Report, 1944, 137th Station Hospital," 11 January 1945, RG 112 Box 123, National Archives; Ferk interviews. According to Eugene Solomon (see below) the fifty or so Army and Navy nurses on Guadalcanal were housed in a small camp surrounded by a 25-foot tall fence with only two gates, both guarded by MPs.

23 Frances Sinclair-Anderson to Shoptaugh, July 2007.

24 William Buckingham, telephone interview with Shoptaugh, March 9, 2007. Buckingham later became a physician and practiced for many years in North Dakota. While on Guadalcanal, he played a small part in the struggle to desegregate the U.S. Army. "We formed a little basketball team at the hospital. A fellow named Hank, who was part of the colored Engineer Unit that served as our fire unit, used to watch us practice. We asked him if he wanted to play and he said yes. Gee, I don't know how we got away with it, but he played with us when we played other units. I suppose we were so bad that no one complained. It is of interest that the fellow that was sort of our team leader was from Kentucky and he made sure Hank played."

25 Eugene Solomon, interview by Leslie Collins, February 24, 2003; Shoptaugh telephone interviews with Solomon, March 17 and April 24, 2007.

26 "Annual Medical Report, 1944, 137th Station Hospital," 11 January 1945, psychiatric illness section.

27 Cooper and Smith, p. 304; Cronin, pp. 213-14.

28 Bougainville pastimes are described in George Dingledy's letters and in the letters of Lt. H. A. Tyl ('HAT') that were published in 'HAT" Tells How Grand Forks And Dickinson Men Spend Their Days And Nights Out On Bougainville Fargo *Forum,* late 1944 (clippings in Russell Opat papers). Emil Blomstrann's memoir is in "Company E, 164th on Bougainville, 1944," *16th Infantry News,* October 2007, supplemented by Shoptaugh interview with Blomstrann. Freeman information from Shoptaugh telephone interview with Warren Freeman, July 17, 2007 and clipping from Springfield Massachusetts newspaper, "Souvenirs from Soldier-Husband," courtesy of Warren Freeman (William's son).

29 Doug Burtell related the story about Eli Dobervich in conversation with Shoptaugh. June Dobervich, Eli's fiancé and wife, told Shoptaugh in 1996, that her husband had "nightmares about the war, for the rest of his life." The drunken officer story is in Walker, p. 77, and several other men interviewed also related stories of Coronel Mahoney's drinking habits.

30 Walker, pp. 85, 141-42, 144; Milton Shedd letter to his wife, August 11 1944. Shedd's letters were complied in a booklet in 1945 and a copy of this was made available to Shoptaugh, courtesy of William Shedd, Milton's son.

[31] Changes on M Company found by comparing company roster from 1941 to that in October 1944 roster, in the papers of John Tuff (courtesy of Edith Tuff).

[32] Edwardson conversation with Shoptaugh, September 2008; Cooper and Smith, p. 304.

[33] Albert Wiest, e-mail to Shoptaugh, November 2009.

[34] Vincent "Swede" Clauson, interview with Don Knudson, January 28, 2007.

[35] John Paulson, "A Patrol on Bougainville," *164ᵗ Infantry News,* December 2000.

[36] "Report of Operations, 164th Infantry, 1 May 1944 to 1 Dec. 1944," p. 5, 164th Infantry Papers,; Pete Baldino Diary, entries for August 14-23, 1944 (copy of diary from Baldino family). A short, undated biographical sketch of Baldino, written by his brother Louis, states that Pete was killed when he "bravely shielded the other men" of his unit from the grenade blast. So far as is known, Baldino was never recommended for a decoration.

Chapter 13: "The Ones Who Slap You Back Are Filipinos."
Leyte: January-March, 1945

Looking back on it, George Dingledy said "you know the funny thing was none of us ever figured it would be over before 1948. That was the honest to God truth. We'd been gone for a couple of years already, but we never thought much of going home until the damn thing was done." As he wrote a friend in 1944, he thought after wrapping up at "island x" then "we'll be moving on up and duplicating our work in some other location." Once the Japanese retreated from the perimeter, the service troops "were sitting there [in the American perimeter] having a good time. I could have spent the rest of the war right there [on Bougainville] without much problem."[1]

After late September there were no combat casualties in the regiment. In November, Milt Shedd led the reconnaissance platoon deep into enemy territory, surrounded a large force of Japanese soldiers and wiped them out. They then returned without a single casualty. "Surprise and correct use of our superior weapons had caused these results," Shedd wrote home. . "The American soldier is tops and knows his stuff." For this "perfect patrol," Shedd was awarded the Silver Star.[2]

The command was shuffled again. Crump Garvin was promoted and Colonel William Mahoney, Garvin's executive officer, took command of the 164th. Bill Considine became Mahoney's exec. John Gossett got command of the 1st Battalion, Samuel Gee the 2nd Battalion and William Mjogdalen retained command of 3rd Battalion. "Officers came and went and we seldom bothered to keep track," one man commented.[3]

The men switched from playing baseball to football, speculated on how long the war in Europe would last, and were entertained by a few more USO troupes. The Americal received several shipments of mutton from Australia. The men hated it, but fortunately found an Australian unit that was willing to trade their beef rations for the mutton. Mostly the men wondered where the Americal Division would be sent next. In December companies were mustered for "pre-amphibious training" -- climbing up and down practice nets. After that, they knew something was coming.[4]

The Japanese were falling back across the Pacific. General MacArthur's forces secured New Guinea and Rabaul was rendered powerless by bombing. With a series of brilliantly executed landings, MacArthur cut off thousands of enemy troops, starved them of supplies and left them to "wither on the vine." An Allied force seized the island of Biak, where engineers built airfields within range of the Japanese base at Davao in the Philippines. The Japanese lacked the forces to prevent the Allied victories. When Admiral Nimitz's invasion fleet seized the Marianas, the Japanese committed their Combined Fleet to a counterattack. But the Battle of the Philippine Sea was so one-sided the American fliers called it the "Turkey Shoot." The Marianas were converted into bases for the new B-29 bombers, and the Japanese home islands came under bombing.[5]

B-24 bombers pummeled Japanese bases in the Philippines. Rolf Slen, the young navigator with the 494th Bomb Group, began flying missions over the Philippines in November 1944 from Angaur, an island in the Palau chain. "We lived in tents on Angaur and slept in cots.

Our tents were pitched practically on the beach. We had a nice lagoon practically outside our tent, but it had some vicious undertows so you had be careful swimming. Then one time a bunch of huge crabs came out of the water and littered the beach, and then suddenly disappeared. We never saw them again." The lagoon was nonetheless welcome, for the temperature on Angaur rose to 115 degrees.[6]

Most of the missions that the 494th flew were against Japanese bases in the Philippines. "We bombed a lot of airfields, barracks and camps, and sometimes tried to hit individual gun emplacements that were no more than fifty feet square." Unlike the large formation-bombing against Germany, the Pacific bombers usually "flew a very loose formation from Angaur to the Philippines. A run on the target had to be straight and could take from one to ten minutes, but whatever it was, to me it seemed too long. I had nothing to do but sit there and hope to hell we'd get out of there pretty soon. I suppose we ran into Japanese fighters four or five times, but never had any damage from those. Most of our opposition was flak. One time a shell went through our wing. Luckily it didn't explode on impact but above the wing. I remember that at times I was sitting there thinking about maybe having to jump out of an [damaged] airplane over the ocean. That was suicide, they'd never find you."[7]

The U.S. Sixth Army already had landed on the island of Leyte and MacArthur broadcasted his message to the people of the Philippines: "I have returned." The Japanese tried to make it a short visit. In order to destroy his naval support, they used the *Kamikaze* suicide planes in great numbers for the first time. The remaining ships of the Imperial Japanese Navy set out in one last major confrontation but the overwhelming predominance of American power proved to be too much. The U.S. Navy sank four carriers, three battleships, ten cruisers and eleven destroyers. Its own losses were negligible by comparison. As one of the Japanese admirals told an American interrogator after the war, "we had to do something and we did our best. It [the Battle of Leyte Gulf] was the last chance we had, although not a very good one."[8]

Once the Americans were securely ashore on Leyte, MacArthur wanted more troops in order to advance to Luzon while continuing the Leyte battle. He asked for the Americal and 37th Infantry divisions to be transferred to his command. In early November, the entire XIV Corps was alerted for shipment to Leyte. MacArthur intended to liberate all of the Philippine islands. This made little military sense, but MacArthur insisted on this campaign in order to erase his defeat in 1942. "He is insane on this subject," General Griswold wrote after the XIV Corps joined MacArthur's command. MacArthur "said he expected little opposition, that the battle of the Philippines had already been won on Leyte. I do not have his optimism." Nor for that matter would the men in the Americal.[9]

The men in the 164th Regiment took their departure from Bougainville with mixed feelings. Despite the heat and wet and malaria, most of them were used to the quiet months since summer. Tents had been dressed out with awnings, hammocks had been strung up, and volley ball courts improved with lights. Men made pets of birds. There was body surfing in the bay with filled mattress covers. Fishing competitions had become common; one officer noted, "not many fish were caught, but countless bottles of beer were consumed." Bob Dodd and his friend Bob Olson talked a supply sergeant into lending them shotguns and then rented trained water buffalo from the natives. Riding the backs of the buffalo, they went out "to shoot teal at an inland lake." No one was in a hurry to get back to war.[10]

Once the word came about the Philippines, the soldiers packed up, discarded excess baggage, and left Empress Augusta Bay in transports between January 8 and 10. They rendezvoused off Hollandia and joined a larger convoy bound for Leyte. For those who had been part of the regiment since the war began, this was their fifth sea voyage since leaving San Francisco. Chuck Walker recalled the "exceptionally heavy" swells that rolled his transport this trip: "at its uttermost slant, hesitating there for as long as thirty seconds, someone said 'She's going over next time.'" But the waves slowly gave way to calmer water and the convoy continued north.

The convoy made port at Tacloban in Leyte. They went almost immediately into action. A substantial number of Japanese infantry still held out in the hills between Carigara Bay to the north and Ormoc Bay to the south (see Map 8). Many men would die in the effort to subdue them. The campaign would be all the more bitter because even before the whole of the Americal Division arrived, MacArthur had announced that the island was secure and all that remained were "minor mopping up" operations. Eager to get on to Luzon, he left no more than 36 small ships behind for supplying the men on Leyte.[11]

Leyte was very different from the Solomons. The climate was still hot, but not so horribly covered with jungle; much land was cultivated in sugar cane, palm trees, rice, cane and bamboo fields, and some varieties of potatoes. Small hills were the most important military feature. A well-placed machine gun on one of these could sweep the broad plains of the northwestern peninsula. The native population relied on caraboa (Filipino water buffalo) for both their diet and their transportation. There were also thousands of lizards and "grey back" bedbugs that appeared to be everywhere, in the trees, villages and fields.[12]

The Filipinos lived in small villages for the most part, usually in wood shacks covered by dried grass and palm. Due to become an independent nation in 1946, the Philippines was still a territory of the United States. Most of the Filipino population hated the Japanese and were very happy to see the Americans. Some of the Filipino women in the village of Valencia were a bit too friendly. "It was apparent these women had associated with troops before, as most were evidently prostitutes," one officer recalled. "I was forced to place a temporary guard to put the run to them." Most everyone on Leyte spoke some English. Filipino children had been orphaned during the Japanese occupation, and the men gave them gifts of clothes and food.

Joe Castagneto and others in the regiment who were Catholic noted how the small churches on Leyte resembled the neighborhood churches they had attended as kids. Some men were worried that they could not easily tell the difference between the enemy and the Filipinos. Both looked "like Japs," they said, so how could they avoid shooting the wrong man? John Gossett told several of them something he had heard from a naval officer: "Line them all up and slap their faces. The ones who slap you back are Filipinos."[13]

On January 28, companies from the 164th were trucked from Ormoc to the *barrio* of Valencia. Their first contact with the enemy came about a mile west of the village of Carigara, when a patrol of Company E met a group of Filipino guerillas who gave them a prisoner. Milt Shedd took the prisoner back to his headquarters and a couple of translators interrogated him. He was happy to talk in return for a meal. He had deserted from his infantry unit and had "been lost for 10 days" while he foraged for something to eat. "Morale of [the] troops is very poor," he told the Americans, "and food is scarce." The veteran soldiers still expected the Japanese would fight. Dick Stevens was now a squad leader; he told his new men to expect any Japanese soldier to fight on until he was killed.[14]

The Americans would have to work with guerilla bands on Leyte. Some Filipino guerilla bands were well led and maintained some semblance to discipline. Others were no better than bandits who preyed on their own countrymen. And it was difficult to tell exactly what guerillas might do when shooting started. Many men wanted the Army to disarm the guerilla groups but MacArthur dismissed all such suggestions.

Lieutenant Charlie Ross, by contrast, admired the zeal of a brash guerilla that joined one of his patrols in late January. The patrol had already flushed out and killed several of the enemy and was now looking for "two [enemy] machine guns on the side of a mountain."

The guerilla, a very young man named Victorio Maget ('Vicky'), showed them the location of the guns. As they moved closer, they came upon a grass shack and 'Vicky' immediately rushed inside, firing his rifle. He killed two Japanese officers and "came out carrying their Samurai swords." The patrol moved on. As they climbed the steep hillside, trying to work their way around the gun positions, Vicky was killed by an enemy sniper. Ross in turn killed the sniper. Bob Jeffries took over the point and got the patrol behind the guns, but the Japanese already had pulled out of their position and disappeared.[15]

Fighting in the Philippines was like the post-March fighting on Bougainville – patrolling and skirmishes with Japanese units that held out until the last minute and then fled, only to turn up entrenched in another position. Tracking them all down was like trying to use a trawler net to catch goldfish. The Japanese had limited supplies of food, no air support, no heavy artillery. But their rifles were still accurate, their machine guns still plentifully supplied with ammunition, and their tenacity was only slightly diminished. There were plenty of chances to get killed in a fight that did little to bring the end of the war closer. In an effort to coordinate small actions for all three regiments, assistant division commander Brigadier General Eugene Ridings set up an advance headquarters alongside the 164th HQ to make "major tactical decisions."[16]

The Japanese intransigence left Americans in little mood to take prisoners. When Charles Walker prepared to take his men forward from Valencia, his battalion commander told him "don't bring in any more prisoners; we have no room for them." Walker asked for that to be in writing, but his boss said "just do it." Walker ignored him, but some of his men were prepared to act on their own: a few days later as Walker's patrol found a shack filled with "several sick Japanese," Walker had himself barely entered the shack before "one of my men behind me cut loose with a BAR, shooting from the hip." Walker rebuked the man with "swear words I hardly ever used."[17]

The intelligence officers 8th Army, to which the Americal was attached, had estimated that about two or three thousand Japanese soldiers were in this part of Leyte. There were actually more than six thousand of them just in the area of the Americal attacks. Over the next weeks, patrols set off from the 1st and 2nd Battalions They flushed out and killed several dozen of the enemy in a variety of skirmishes, and pushed them back to the coast, usually suffering very few killed but one or two men wounded. While they had a decisive advantage in fire support, tanks were of limited use in the terrain and the fighting was soldier to soldier. It was a hell of a thing to ask a man to risk his life for 'mopping up.'

Horace Nearhood had patrol experience from Bougainville. "On my very first, January 1, 1944, our scout got shot and died while we carried him back. That grows you up in a hurry." Neither Horace nor his twin brother Forrest had been hurt on Bougainville. But now, in the Philippines, as B Company moved west from Valencia, the odds caught up with Horace. "I think we'd been there only two weeks, and it was like our first day out in these hills. We spotted a Jap patrol running along a long sloping hill. We started up and got about half way up the hill when our Sergeant called to us to come back. They were going to put mortar rounds on the hill first.

So as they started down I stayed back as BAR man to cover them. I crouched into a small depression with a lot of gravel around it. It was then a bullet hit the gravel in front of me, bounced up and creased my forehead and it knocked me out." The bullet ricocheted into his shoulder. "Sergeant Kjera ran up and pulled me back."

Nearhood's head wound looked serious. His face was covered with blood. "I couldn't see and the medic thought I'd lost an eye." Fighting broke out and he spent the night on the hillside. "I was full of morphine and kind of out of it. I kept saying I wanted to talk to my brother. A major had come up, I don't know who, and he thought I was delirious, but Sergeant Kjera told him my twin brother was in another squad. The major I guess found Forrest and sent him over to talk to me. He was mad, cussing at me for getting wounded and saying there was nothing worth fighting for here. I think I told him to write the folks and tell him I'd be okay. The next morning the squad carried me back." Once the bleeding from his creased forehead was stopped he could see. Doctors removed most of the fragmented bullet in his shoulder. "I still have the bullet at home and a bit of it is still buried in my shoulder, which is why I have forty percent disability at the VA." Nearhood spent several weeks in a hospital on Biak. Then he set off on a long flight for the States. "I was in stateside hospitals until September of '45, then got a thirty day leave and was then discharged." His brother Forrest came home after the war without injury.[18]

The tent hospitals on Leyte treated the lightly wounded and evacuated the serious cases to hospital ships. It was a tossup as to whether the hospital ships were safe. Japanese *kamikaze* pilots did not flinch at attacking these vessels. The American air forces shot down dozens of *kamikazes* and sank numerous small boats the enemy used to send reinforcements to Leyte[19]

The 164th's 3rd Battalion made a sea landing to seize the village of Villaba, on the west coast of the peninsula. Mahoney hoped this maneuver would trap and destroy the Japanese. The 3rd Battalion had expected to meet a group of Filipino guerillas at Villaba, but for reasons that were never explained the guerillas already had left the area. As a result the enemy was able to contest the beach. When the men in Company K approached the Villaba pier, a machine gun opened up, Private Paul Guzie was in one of the boats. He had joined the 164th on Bougainville, and credited the "Guadalcanal vets with showing me how to keep alive" while on patrols. He "encounter[ed] Japs that led to a minor skirmish" only once on Bougainville. Now at Villaba "we were met with machine-gun fire on the beach. One man got hit in his rear end. I often wondered what he told his friends about how he got his purple heart. Another man had a bullet hit him across his chest, cutting the bottoms of his chest pockets. Two grenades fell out and fortunately did not explode." The men had to fight their way into the village. The Americans soon silenced the enemy gun with automatic weapons and grenades.

The company's full complement of riflemen was ashore by mid-afternoon. They cleaned out the remaining Japanese and dug in on the high ground outside the village. That night the Japanese counterattacked. Guzie had set up his machine gun "in a foxhole on the east side of the hill. Another machine gun was in the foxhole next to me. We had set up our guns to traverse about three feet [and sweep] about 10 inches off the ground.

Our fire could kill anyone crawling or walking up the hill. During the night I could hear my partner swearing because his gun jammed and he could not get it to fire." Later the man found that an enemy bullet had actually hit his gun barrel. Guzie kept firing until the attack ended. In the morning he found a dead Japanese officer "about two feet from my gun. He had pushed his sword into the ground just inches from my foxhole." Guzie was now a blooded veteran.[20]

K Company now pushed out in pursuit of the enemy. Tony Czarnecki moved up with his machine gun squad, feeling certain that his new platoon leader was going to get some of them killed. "On Bougainville we had a good leader, Sergeant Tony Wolf. He went home after we got to Leyte so we had another guy and he was lousy. If anyone should have been shot for incompetence, it was him." The sergeant told Czarnecki to set up his gun "on the berm of a drainage ditch, where it was sure to draw fire." Czarnecki started to argue with him but this was interrupted when an enemy machine gun opened fire. "Jaime, our squad leader, tried to spot it but got hit in the buttocks. I pulled him back into the ditch by the ankles but he was hit twice more before we got him in. That jerk of a platoon leader was keeping his head buried the whole time. Jaime survived and I took over the squad. I argued with that fool several more times and he asked the captain to court martial me. Nothing came of it. I still regret never punching that guy."[21]

Marion Minks was one of the newcomers to Company K. Minks was working in a defense plant in Indiana when he was drafted in 1944. "I had a wife and a son, and another son was on the way when I finished basic training and went overseas." After some brief jungle orientation in New Guinea, Minks and dozens of others went on to Tacloban. Assigned briefly to a replacement depot, he quickly went into the 164th's Third Battalion. "This was a proud outfit from its time in Guadalcanal. I was a rifleman with K Company. We were told we'd be 'mopping up.' Well, it seemed to us like every hill had a number and a fight. They would attack us at night. With no flares we just fired at movements, and find their bodies near the rifle pits every morning. Near Villaba we started up one hill and they shelled us. We went on, firing as we climbed. Near the top I was moving along near a friend who was a medic. A grenade went off and it got us both. He was wounded and I had fragments in my chest and right leg. It was almost unbearable for me to breathe. I didn't know that the blast had deflated my left lung. We got carried back and I'm sorry I never got to thank those who took me back." Minks was in the hospital until the war was nearly over. "My medic friend was okay too. I saw him once again, about five years after the war. It was good to know he was alright. But I never saw him again after that."[22]

As K Company fought out from Villaba, patrols from the other two battalions were pressing in from Valencia and Libangao. Units from the 182nd were also moving in to encircle the enemy. Some Japanese were falling back while others fought fiercely from their positions. Patrols reported that several of the Japanese seemed "poorly trained" in combat and were easily killed. But while hopes rose for bagging an entire enemy force, they were scattered widely and small groups slipped away. Between February 3 and February 12, the intelligence section of the regiment recorded dozens of short encounters. In one an enemy machine gun fired on a truck convoy; by the time the Americans had deployed and closed on the position the gunners were gone. In another, Charlie Ross led a platoon toward "small arms and 60mm mortar fire" they heard outside the village of Cananga. When they got there they found a mortar unit of the 182nd which were returning fire against snipers in some trees. But the Japanese were already long gone. Soon after that, men from Company A killed two enemy infantry they found hiding in a hut and went looking for more that locals had reported "scattered throughout [the] area." All they found were "6 old dead Japs." Filipinos frequently told GIs that they had seen forty or fifty or a hundred enemy soldiers, but the Americans seldom found more than a few.[23]

    Vince "Swede" Clauson, however, ran into stiff resistance. "We went out on a combat patrol and the Japs caught us by surprise." After a few Japanese showed themselves and lured them into a small valley, others opened fire from hiding and "really pinned us down." Soon half a dozen men were casualties. Swede's sergeant called him forward to lay down cover fire. He came up with his .30 caliber gun, mounted on a bipod. One of his loaders came forward with extra belts, but the third man on his team hung back, claiming he had blacked out after taking cover; "he was a greenhorn and lost his nerve." Swede recalled reading somewhere that machine gunners had life spans of less than a minute. "I decided to have my eleven seconds, and let go real good," firing a belt off in just a few bursts. He used up a second belt and then a third. "By now I was carried away, didn't even think of them as human. I was shouting 'take that you sons of guns' but it was worse than that." The heavy volume of fire forced the enemy down. But as riflemen closed in with grenades, the Japanese scurried away. Swede joined others in carrying the wounded, including a man who "was already dead." One thought kept going through his head: "what a waste war is."[24]

Doug Burtell took part in a reconnaissance patrol sent out to check on a report that the Japanese had a hospital further north in the peninsula. "This was supposed to be near a lake and about ten or eleven of us went to find it. Milt Shedd led the patrol and we had a Filipino guide along. We went along an old trail that had pretty dense jungle around it, and we got to a clearing. You never just went across a clearing because that was a good way to get killed. I went up a little closer and saw a Jap rifle against a tree and behind it there was a guy bending down. I drew a bead on him, but Shedd whispered to me to wait. He turned out to be a Filipino with a Jap rifle and I nearly killed him. We went on to find the lake, but the grass shacks there were abandoned. There were just a lot of dead Japs."

Returning the way they had come, they came on another grass shack. "A group of Japanese soldiers were near the shack gathering coconuts. "Milt Shedd sent me to cross past them and check out the other side." Moving carefully past the shack, Burtell saw another Japanese soldier. "He was coming up toward the shack, dragging his rifle along by the barrel. He didn't see me." The Japanese went on into the shack and soon the other Japanese, their arms laden with coconuts, returned as well. Seeing no others, Burtell waved to the rest of the platoon to join his. "We were now on a little rise of land looking down at the shack and all of us got into a line. When Milt dropped his arm we fired about six clips apiece into that shack. Chunks of bamboo were flying all over from the shack." One man tried to run away but Burtell killed him. "Then we went up and tossed a few grenades in. There were seven dead in there and all they had there was a rusted pistol and a couple of rifles. In a way it was murder, but that was war." Burtell gave his cousin Elroy Greuel a small Japanese flag that he took from the shack.[25]

By the last week of February, the regiment had squeezed the remaining Japanese troops into a shrinking box east of the village of Abijao. Mortars and supporting artillery began pounding the box. Other parts of the 164th, together with the 132nd and 182nd regiments, continued aggressive patrolling to find and wipe out small groups of enemy stragglers. These actions, the divisional history summarized, "varied from intense, bitter fights to boring, uneventful patrols." The battles were fought in "hot, sticky weather and heavy tropical rains [that] plagued the men." Many times the Japanese killed themselves once spotted. Sometimes they fought desperately. On the evening of February 27, Japanese soldiers that were still holding out on a hill that Paul Guzie's platoon occupied suddenly came out and attacked the Americans. Guzie was wounded by an exploding grenade, the fragments lodging in his neck. He later said it felt like "being struck by a large pillow," no pain but a growing numbness. "Just before I was hit, Sgt. [Philip] Dockter from North Dakota was hit and started screaming for help. Our platoon leader tried to calm him down by telling him he was alright. The Sgt. swore and yelled back, 'God damit, I know when I'm dying.' He was right. He died about ten minutes later." Dockter had fought several times in the Solomons, but on Leyte his luck ran out.

Guzie and his comrades drove the Japanese back, killing at least ten of them. After a medic bandaged his wounds, he was able to walk back to an aid station where two doctors removed all but one fragment from his neck. The last was too close to a main artery. He met a lot of wounded men in the hospital and "could see some of them would be crippled for life." When he was released and told to return to his unit, an administrative officer located transportation for him while he spent a few hours "eating and drinking at a PT boat base." Then it was back to the war.[26]

The Americal's loses overall were relatively light and by the beginning of March over two thousand enemy combatants in the Leyte peninsula had been eliminated. The 'kill ratio' was over twenty to one. It seemed likely that with another major push, the Americal Division would be able to eliminate the remaining pockets of Japanese in this part of Leyte. But on March 8 the division headquarters ordered all three regiments to pull back to Ormoc and Vallencia. 1st and 2nd Battalions waited for new orders while the 3rd Battalion was placed in 8th Army reserve. Colonel Mahoney issued a commendation congratulating the regiment for "fine accomplishment of a most difficult mission." Everyone was pleased but knew another mission for them was in preparation.[27]

Meanwhile a few more men were able to go home. The Army introduced a system of points that men 'earned' according to the number of months they had been in the service, the number of campaigns they had been in, medals they had received, and whether or not they had children. A man with 85 or more points would be eligible to go home. Bernard Scheer was dropping 60mm rounds on "some Jap's gun position" when a runner came up to tell him to report back to headquarters. "I dropped one more round into our 60mm tube" and went off to the company headquarters where he met Lester Bethke, his cousin and another old hand in the regiment. "What was up?" he asked. "I guess we're going home on rotation," Bethke replied, adding "I just about didn't end up being here today." He explained that the night before he had fired in "pitch dark" at a shadow he saw near his foxhole. At first light he found the dead Japanese. Bethke then said, "You and I are lucky that we are getting out of here because I think we are stretching our luck a little." The two men sailed home on a Liberty ship four days after that.[28]

MacArthur, despite opinions in Washington that it was not necessary to liberate the entire Philippine archipelago, gave orders to his 8th Army commander, General Robert L. Eichelberger, to schedule landings on the islands of Panay, Cebu, Negros, and Bohol. These islands made up the heart of the Visayas, due west of Leyte and between Luzon to the north and Mindanao to the south. Before the war the Visayas were sources of food for all the islands and also of cane sugar for the U.S. Guerillas already had driven the Japanese into the towns on the islands and MacArthur wanted them completely cleared. He stated that he wanted the islands' agricultural lands in operation and to use it for stationing troops prior to an invasion of Japan. The invasions began on March 16 with landings on Panay by elements of the 40th Division. The fighting there was fairly brief, but nine days later, when the same troops landed on the northern shores of Negros, "hard fighting" developed. It ultimately cost the division almost 1400 men, killed and wounded, to take control of half of Negros.[29]

The enemy on southern Negros, Bohol and Cebu, were earmarked as the Americal Division. The plan for the landing at Cebu called for units of both the 132nd and 182nd to land at Talisay, about five miles south of Cebu City. After securing a beachhead, the two regiments would move quickly to seize Cebu City and the enemy airfield just beyond it. Americal's divisional intelligence men estimated the Japanese force on the island at about twelve thousand men, made up largely from units of the 102nd Division, some naval troops and scattered groups of refugees from Leyte. Filipino guerillas on Cebu scouted the beaches and assured the Americal's G2 men that the beaches would not be heavily defended. They warned that the Japanese had mined parts of the beach. They had not learned that the Japanese had used not only standard land mines but also buried aerial bombs under the surf. Navy survey teams went in at night to map the mines but missed many of the aerial bombs.[30]

On the early morning March 26, a sixty ship task group crossed the Bohol Straight carrying the 182nd and 132nd regiments from Leyte toward the beach. The seas were very calm and the crossing was made without incident. Meanwhile underwater demolition men crept into the bay off Talisay and began clearing any obstructions offshore. After dismantling large obstacles that the Japanese had placed in the surf, the demo men found some of the bombs buried in the north section of the beach, the section where the lead companies of the 132nd would come ashore. "Word concerning the mines," an after action report later stated, were "passed over the radio, but it apparently had not reached these troops." It was only "very good fortune that none of our men had stepped on a mine," the report noted. The soldiers were not going to be so lucky.[31]

The task group hove to offshore while destroyers shelled a two mile stretch of the beach. B-24s and Army dive bombers from Leyte flew in to bomb suspected enemy sites while fighters overhead stood by to prevent a Japanese air attack. At 8:30 a.m. the first wave of landing craft was approaching the shore, while special craft with rocket launchers blanketed the beach with one more round of high explosives. Still no one in the task group realized there were mines buried right where the leading wave would charge out of the surf. William Krentz was a corporal with I Company of the 132nd Regiment. He recalled how lovely the island looked. "The sky was clear, the sea very calm, there was a freshness in the air typical only of the Pacific islands." The barrage that hit the beach impressed him: "for two miles a furnace of flaming geysers of smoke, sand, rocks, trees, and 'what have you' rose from the beach. Even 1,000 yards away the noise was deafening." Men said "nothing, absolutely nothing, could live through that." Soon the LCIs closed to the shore and Krentz heard the order to "drop life belts, load and lock weapons, fix bayonets." Then he "felt our boat ride over the last breaker and scrape the sand

– the ramp was down, [our] machine guns stuttered on all sides." There appeared to be no answering fire as the men rushed across the sand. Then those ahead of him "ran into a solid field of mines and were shredded in a murderous hail of exploding shrapnel." The coxswain in charge of Krentz's craft panicked and put his engine in reverse. Krentz and several other men ran out into deeper water, forcing them to struggle to the shore, laden down with their equipment. He reached the sand feeling "used up," gagging from the salt water in his throat.

Krentz saw mines "bursting everywhere" and within minutes knew that some Japanese had survived the barrage. Fire from what he thought was a "20mm anti-aircraft gun" ripped into the ground near him. "A mortar man on my left was disemboweled; the blast knocked me into a shell hole. I felt warm urine running down my legs." Krentz's sergeant yelled at him to get up and move in, so he clambered out of the hole, passed by the dead man whose "torso was still smoking," and headed in, praying that he would not hit a mine. After making it about a hundred yards past the water, he and others spotted the enemy gunner "barricaded in a bamboo shack." One man pushed a mortar's base plate ahead of him while crawling up to the shack. Then he kept the gunner pinned down while another GI incinerated the shack with a flamethrower.[32]

Along the beach, casualties from the land mines writhed in the sand. Many others who had not been injured were in a daze, fearful of moving further into the minefield. Smoke added to the confusion. Two thirds of the fifteen landing craft in this part of the first wave had been put out of action by the explosions and gunfire. Casualties from the crews were high. The second and third waves were held up offshore, in some cases unable to get past the wreckage, in others their coxswains hesitate to go in. General Ridings found it necessary to commandeer a boat to take charge. The coxswain of his craft managed to push past the chaos and get him to the beach. There he proceeded to get men busy "probing for and taping routes through the obstacles." By 1000 he had managed to get the beach "completely unjammed" and the advance back on track. The men moved inland and settled down for the night while scouts checked the surrounding terrain.[33]

On the 27th the soldiers set off for Cebu City. A number of them were still nervous from the bloody beach, and they worried what might await them in the city's narrow streets. Corporal Krentz was surprised that no enemy attacked them on the six mile march. As his unit entered the outskirts, he was told that only a small "skeleton rear guard" had been found in the city. He noted that Japanese "helmets hung on the rifles" of the advance men. Moving on into Cebu City, Krentz noted that many buildings were heavily damaged and the streets all but deserted. Soon, however, civilians began to appear from hiding and cheer the men as liberators. One of these, a man named Melero, who was from the Spanish embassy to the Philippines, produced bottles of scotch whiskey. For a brief moment Krentz and his friends enjoyed the privileges of being conquerors.[34]

The Japanese commander on Cebu, General Takeo Manjome, had decided to abandon Cebu City. Many of his men had no combat experience; some indeed were recently drafted Korean laborers and civilians from Japan.

Manjome chose to make his stand in the hills north of the city. His men had converted caves into fortresses and hill tops into a system of entrenched pillboxes with interlocking fields of fire. Manjome had no illusions that he could win the coming fight. He just wanted sell his men for as many American lives as possible. His strategy began to pay dividends as soon as the lead companies of the 182nd advanced toward the airfield complex north of the city. They walked into a hail of machine gun fire from the hills ringing the airbase and casualties mounted up quickly. Companies from the 1st Battalion of the Bay Stater Regiment reorganized to clear two of the hills. But as they were gaining ground on one of these, a hidden enemy triggered a booby trap that set off a pile of explosives and ammunition. The explosions rocked the ground and chopped down dozens of men.

The fight for the hills was just beginning. Between 28 March, when the battle began in earnest, and 12 April, both the 132nd and the 182nd threw every man they had "head on" at Manjome's hill defenses. Ridings complained that he did not have enough attacking forces to pin Manjome's men in place and then flank his line. Both the 132nd and 182nd were under strength and the men were tired from the Leyte fighting, he said. LSTs converted into temporary hospitals hauled hundreds of wounded men out of the harbor at Cebu City and took them back to Leyte. General Arnold reviewed the loss reports and came to the conclusion that Ridings needed help. He sent a message to the 8th Army headquarters asking for the reserves. General Eichenberger agreed and Colonel Mahoney was ordered to get the 164th ready for transport to Cebu. Its short rest was over.[35]

(Endnotes)

[1] George Dingledy, letter to "Francis, Jeanne and Michael," February 28, 1944, and interview with Shoptaugh, September 14, 2007.

[2] Shedd's patrol is described in his letters and in his medal citation, both published in the *164th Infantry News*, October 2009.

[3] Years later, a retired Garvin told interviewers that the 164th was "superior to [all] the Regular Army units he had commanded in nearly every respect" – Cooper and Smith, *Citizens as Soldiers*, p. 329. Mahoney would remain in command of the regiment until the end of the war. Burtell comment from conversation with Shoptaugh.

[4] "History of the 164th Infantry, 1 April to 30 June, 1944," p. 260, 164th Infantry Association Records; pre-amphibious training in Cronin, *Under the Southern Cross*, p. 216.

[5] Morison, *New Guinea and the Marianas,* pp. 45-147, 260-61. Morison credits the American carrier air intercept system for the destruction of the Japanese air strikes. See also Robert Ross Smith, *The Approach to the Philippines,* pp. 280-449. For an incisive look at the conduct of the New Guinea campaign, see Stephen Taaffe, *Macarthur's Jungle Victory: The 1944 New Guinea Campaign* (Lawrence, KS: University Press of Kansas, 1998).

[6] Rolf Slen, interview with Shoptaugh, July 29, 2007. The occupation of Anguar as part of the Palau campaign is discussed in Robert R. Smith, *The Approach to the Philippines,* pp. 494-531, while the construction of the Angaur airfield is described in Karl C. Dod, *The Corps of Engineers: The War Against Japan* (Washington DC: Department of the Army, 1966), pp. 514-15. The much bloodier battle for the neighboring island of Pelelieu is well covered in Bill D. Ross, *Peleliu: Tragic Triumph* (New York: Random House, 1991).

[7] Slen interview, supplemented by Slen, "A Memoir of World War II," p. 20. For a good account of the 7th Air Force, consult Clive Howard and Joe Whitley, *One Damned Island After Another* (Chapel Hill: University of North Carolina Press, 1946); *George Kenney's General Kenney Reports* (Washington DC: Office of Air Force History, 1987 reprint of 1949 edition) contains much useful information on the 5th Air Force. Both books tend to exaggerate bombing results.

[8] Thomas J. Cutler, *The Battle of Leyte Gulf, 23-26 October 1944* (New York: HarperCollins, 1994) contains information not available in earlier works on the subject. The Japanese admiral's remark is from the interrogation of Vice Admiral Takeo Kurita, October 1945 (INTERROGATION NAV. NO. 9, US Strategic Bombing Survey interrogations, No. 47), available at http://www.ibiblio.org/hyperwar/AAF/USSBS/IJO/IJO-9.html.

[9] The American departure from Bougainville and the subsequent Australian experience is discussed in Gailey, pp. 191-211. Smith *Approach to the Philippines*, points out that MacArthur's troops in Luzon had more difficulty capturing Manila because so many of his forces were tied down in Cebu and Bohol, while Ronald H. Spector, *The Eagle Against the Sun: The American War With Japan* (New York: Free Press, 1985), terms the Visayan campaign of "no strategic value whatsoever."

[10] Charles Walker, *Combat Officer*, p. 140; Robert Dodd, Jr., *Once a Soldier*, p. 78. MacArthur's tendency to dismiss "mopping up" operations as unimportant was universally resented. The Australian War Memorial Museum keeps a poem written by an unknown veteran of the New Guinea fighting, with the lines: "They drained to the dregs hell's cup; But the blood they gave was a trifling thing-They were only **"mopping up."** (Emphasis in the original), see the Free Post web site (http://www.freerepublic.com/focus/f-news/953390/posts), posted July 27, 2003.

[11] Cooper and Smith, *Citizens as Soldiers*, pp. 306-307; Cronin, pp. 224-25; Walker, p. 147. M. Hamlin Cannon, *Leyte: The Return to the Philippines* (Washington DC: Department of the Army, 1954), is a very good official account of the battles on Leyte up to mid-January 1945, but summarizes the fighting after that in a mere ten pages, noting that the battles fought by the American and other units 8th Army were "dangerous, difficult, and unglamorous" but "highly essential" to the island's security, p. 365.

[12] Conlin, *Leyte*, has good descriptions of Leyte, while the observations about bedbugs and lizards are from Charles Walker's interview with Blake Kerbaugh.

[13] Walker, pp. 149-50; Castagneto interview with Shoptaugh.

[14] S2 Journal entry for January 28, 1945, 164th Infantry Association Records; Stevens telephone interview with Shoptaugh.

[15] Walker, pp, 152-53; Charles Ross, "A Tribute to Platoon Sergeant Robert E. Jeffrey," *164th Infantry News*, March 2002.

[16] Cronin, p. 234, explains the Gee-Rider supervision of the Leyte actions. The best summary of the major actions fought by the 164th on Leyte is the S3 "Chronological Report of Battalion Operations from 11 February to 11 March, 1945, copy in the 164th Infantry Association Papers. The daily S2 reports provide some sense of the frustrations involved in chasing down small groups of enemy defenders, with frequent remarks like "little evidence of casualties," "unobserved results," etc.

[17] Walker, pp. 154-55. Americal Division commander William H. Arnold admitted in his postwar interview that he did not encourage the taking of

prisoners – "the average Japanese prisoner knew nothing whatever about anything," – and that American and Japanese soldiers would both "shoot people who would probably have surrendered." See Hastings, *Retribution,* p. 186.

[18] Horace Nearhood interview with Shoptaugh, supplemented by Nearhood interview with James Fenelon. PCF Robert Love, another machine gunner with B Company, believed that the "twin" had been killed and could "still hear that brother screaming about it all night" sixty years later. Love telephone interview with Shoptaugh, January 23, 2008.

[19] Field Order 2, February 8, 1945, 164th Infantry Association Records; "History of 164th Infantry," 4 January – 5 April, 1945," pp. 496-97. For medical care on Leyte, see Albert Cowdrey, *Fighting for Life,* pp. 296-99. Samuel Eliot Morison, *Leyte* (Boston: Little Brown, 1958) pp. 395- 396, notes the destruction of Japanese boats trying to reach Cebu.

[20] Paul Guzie, "From New Caledonia to Japan, 1943-1946," 2006, together with Shoptaugh interview of Guzie, July 7, 2007. See also Cronin, p. 229-231; Cooper and Smith, pp. 307-8.

[21] Czarnecki, telephone interview with Shoptaugh, August 22, 2007. I believe that Tony Czarnecki manned the gun damaged by the Japanese bullet in the story related above by Paul Guzie. Czarnecki told essentially the same story, describing the same gun damage in his interview.

[22] Marion Minks, telephone interview with Shoptaugh, February 8, 2008.

[23] S2 Journal entries for February 3, 1945, 164th Infantry Association Records.

[24] Vincent "Swede" Clauson, interview with Don Knudson, January 28, 2007.

[25] Burtell interview with Shoptaugh, October 12, 2007, with additional detail from Burtell interview with Robert Dodd Jr.

[26] "History of 164th Infantry," 4 January – 5 April, 1945," and S2 Journal entry for February 27, 1945, both in 164th Infantry Association Records; Paul Guzie, "From New Caledonia to Japan, 1943-1946," together with Shoptaugh interview of Guzie, July 7, 2007.

[27] Cooper and Smith, pp. 309-13; Cronin, p. 263. Mahoney's March 30 commendation is in the March 1988 issue of the *164th Infantry News*.

[28] Bernard Scheer, "Resume of My Army Career." For how the system worked, see Geoffrey Perret, *There's A War to Be Won,* chapter 30.

[29] Robert Ross Smith, *Triumph in the Philippines* (Washington DC: Department of the Army, 1963), pp. 601-17. Cronin, pp. 265-66 and Samuel Eliot Morison *Liberation of the Philippines* (Boston: Atlantic, Little, Brown, 1961), pp. 228-33, summarize the Panay and northern Negros operations. Steven Trent Smith, *The Rescue* (New York: John Wiley and Sons, 2001) has some nice detail on life in the Visayas during the war.

[30] For the Cebu landing plans see Robert Ross Smith, *Triumph in the Philippines*, pp. 608-10, and Cronin, pp. 269-72.

[31] "Action Report, Naval Beach Party Number 5 – Cebu Operation, 12 March to 6 April, 1945," Edwin Nils Johnson Papers, East Carolina Manuscript Collections, Joyner Library, East Carolina University. I am indebted to the special collections staff of Joyner Library for locating this document, preserved by Johnson, a member of Beach Party Number 5.

[32] Krentz's account of the landing is in *Orchids in the Mud*, p. 355-57. See also Cooper and Smith, pp. 313-14, and Cronin, pp. 273-77.

[33] Smith, *Triumph in the Philippines*, p. 612.

[34] *Orchids in the Mud*, pp, 361-62. Cronin, p. 281, notes that some of the damage in Cebu City was the result of the Japanese conquest in 1942, some was from American air strikes during the invasion of Leyte, and the rest was due to Japanese troops who looted the city before fleeing.

[35] Cronin, pp. 284-91, relates the woes of the 182nd and 132nd in battering at the hills north of Cebu City. Smith, *Triumph in the Philippines,* pp. 614-15, reviews the decision to call for reinforcements.

Chapter 14: Into the Breach Again, and Again
Cebu-Negros: April-May, 1945

The 164th landed at Cebu City on April 10. General Arnold admitted after the war that he thought that with the Americal's full complement in the field, he would "have those babies" [the Japanese] in short order. But he quickly learned that it take more time, and lives, than he expected, because the 8th Army suddenly ordered that the 3rd Battalion be used for an immediate landing on the neighboring island of Bohol, evidently believing that this would prevent the enemy forces on that island from slipping across the strait and reinforcing Manjome.

The 3rd Battalion prepared to invade Bohol and the 1st and 2nd Battalions marched up the valley of the Mananga River on Cebu, hoping to flank the entrenched Japanese defenses in the hills (see Map 9). The 132nd and 182nd would pin Manjome in place while the 164th took the enemy in the rear. This was similar to the plan at Koli Point back in 1942. Kola Point had failed to bag all the Japanese, resulting in arguments among the commanders and the relief of a well-respected battalion leader. Would a similar plan work better this time?[1]

The flank march opened on the 11th with an encouraging note. Foliage along the river appeared to shield the companies from enemy observation. There was brief hope that they might surprise the Japanese. But as Company E resumed its route northward, Charles Walker and his men looked over to the eastern ridge and saw "Japanese standing on [the] peaks, peering down on us." Soon after, Walker met up with John Gossett, who had gone ahead and scouted the enemy positions, dressed in Filipino civilian clothes. Gossett told Walker to expect "a tough climb [into the hills] and a tougher fight, since they can see you all the way." You'll have to "dig them out, Gossett concluded; "I suspect it will also be costly."[2]

The Japanese were already waiting for the 164th. On the afternoon of the 13th, as the companies reached a hill complex dominated by what locals called "Mount Babag," mortar rounds began to fall in their midst, wounding several men. The forward observers for each battalion warned that the enemy had a "strong outpost line" and behind it a "trench system . . . with mortar positions and machine gun positions." Surprise was lost and facing such formidable positions, some officers suggested another flank march to the left. After the war General Arnold said that he would have done this but could not because of the 3rd Battalion's manpower being lost to the Bohol invasion. The original plan now called for the 164th to launch a coordinated attack with the 182nd. Apparently prodded by his superiors, Mahoney decided the attack had to go forward.[3]

Division headquarters also ordered the 132nd and 182nd to attack hill positions north of Cebu City, and further decided that the 164th should attack without a preliminary artillery bombardment. Perhaps with a surprise, the Japanese defenses, hit from two sides, would crack. This required very intricate timing. Even if it worked it was still a frontal assault against the enemy's entrenchments. It sounded to the men like something right out the trenches of France in 1918.[4]

In the early hours of April 14, the 1st Battalion moved out on the left, aiming at a rocky hill just in front of Babag, while the 2nd Battalion advanced on another hill to the right. The men made every effort to move quietly, which was wasted effort because, as the divisional historian noted, the Japanese were "fully aware of both the presence and the intentions of the [164th] units." The 1st Battalion's objective was lightly defended and was taken readily. Even then, their flanks were under constant fire from machine guns on other hills.[5]

Companies E, F and G in the 2nd Battalion had it rougher. The men in G Company rushed the hill to its front, took casualties from "intensive" enemy fire and had to fall back. The men of F took their assigned objective but spent the rest of the day under "intense knee mortar fire from points unknown." They lost seven men. Easy Company's men saw no enemy at their point on the hill, but as Walker noted "every inch was covered by frontal and flanking fire from a distance." Casualties began to mount quickly. Walker ran over to one man and found he was dead with a bullet in his head. Seconds later, Gordon Landvik, a veteran of two years, took a bullet in the chest and died in a few minutes. Others went down but the riflemen dug in on the crest of the hill. Fire continued to harass them. Platoon leaders called for mortar fire on suspected enemy positions.

Veon McConnell, who was now the 2nd Battalion commander, came up to the front to see the situation. Walker and McConnell disliked one another, had clashed during the Koli Point maneuver on Guadalcanal, and had argued over most things since. McConnell regarded Walker as an insubordinate know-it-all, while Walker thought McConnell avoided the front lines and "knew nothing about using the weapons company for support."

McConnell had been ordered by Mahoney to go forward on this occasion. On this occasion the two men "mentally declared a truce." Easy Company clung to the hilltop. Men could not move without drawing a mortar shell or machine gun burst. Scouts estimated that at least a hundred and fifty enemy soldiers were deeply dug in on the surrounding hills. The 132nd and 182nd made some progress by "exposing themselves to sudden death" and knocking out pillboxes with flamethrowers.[6]

The next day the infantryman's nightmare continued. To the north, A Company made a rare night advance to grab a piece of the next hill to its front, but was isolated and by mid-morning had to fall back under heavy fire. B and C companies held their ground under fire all day. The 2nd battalion fared no better. G Company's riflemen rushed their objective again, machine guns raking them as they pushed up the slope. The men fought their way to the crest. Enemy soldiers armed with rifles and grenade launchers contested every inch of ground. Assisted by 81mm mortar support and direct fire from anti-tank guns, G Company destroyed four pillboxes and a machine gun nest. But by the evening they had suffered over forty casualties, close to a third of their already reduced strength.[7]

In two days and close to eighty casualties, the regiment advanced no more than two hundred yards. The rest of Babag Ridge was still firmly in enemy hands. Japanese loses so far amounted to ten confirmed killed and one prisoner, but Wendell Wichmann, the regimental S3, thought the Japanese were weakening, as did his counterparts at the 132nd and 182nd. They were right. Manjome began a partial withdrawal northward as soon as the 164th's flank movement was spotted. He needed five days to save most of his force and some supplies; his men on the hilltops were buying him time.[8]

The Americans, tired, battered and bleeding, hauled themselves up and attacked again, and again. Company F made a summary of each day: *14 April* – "Contacted enemy at 22.9-35.6 [the objective hill] at 1100-rifle fire- One man WIA . . . Positions dug on crest Hill #2 – at 1300 one KIA S/Sgt Frank Bohm-Three WIA due to enemy machine gun . . . About 26 [mortar] rounds were dropped including 5 duds. 2100-Three KIA-Capt John Landdeck [commander of F Company] died instantly-Pfc William Fox died at 0200-Sgt Albert Dietl instant death." *15 April* – "Casualties cleared from area by 0830. C Co. assaulted [adjacent] Hill #6 taking 31 casualties and failed to take Hill. Pvt Gomez of Med. Det[achment] with H Co. creased on right side of head with shrapnel 1230. From 0900-1300 crossfire from machine gun, sniper and mortar harasses our position." *16 April* – "Co. still dug in on Hill #2. – 2400 to 0500 all quiet. 0500 activity heard front of left flank position, one enemy seen and fired upon, results unobserved. 0840 B Co. moved off Hill #2 – Lt. Wolfe with one platoon extended our perimeter to cover B Co's former positions. E Co. and G Co. on our right flank. E Co. blew out several pillboxes also artillery got some enemy. Rounds of enemy 90mm landed in Co. area KIA instantly Pfc Debolt, WIA Lt McGee from concussion. 1st Bn took high hill on our left flank at 1800." Fox Company had just 97 men left by the end of the day.[9]

April 16th proved to be the critical day. Manjome was withdrawing the bulk of his men. G Company, bloody but unbowed, went up its hill and stayed, reducing pillboxes and counting 36 dead enemy. B Company "jumped off for [an] attack on the ridge system." Supported by dive bombers, they reached their objective. A Company advanced, too. In 2nd Battalion, G and E (164th) Companies pushed ahead. Chuck Walker shot a Japanese sniper with his M1. "I saw his helmet tumbling backward up the hill. Mac said 'I think you got him in the forehead.'" They saw no other Japanese. Soon after this, B Company made contact with a patrol from the 132nd and a patrol from the 182nd which had cleared the ridges and were pursuing Manjome.[10]

Manjome eluded the Americal division with most of his men. There was at least one case where an advancing platoon from the 164th reached a road in time to see a group of trucks passing by. "Say, weren't those Japs?" one man asked another. Patrols found huge stashes of loot in caves around Mount Babag. Joe Castegneto explored a cave filled with blankets, furniture, clothes, "all kinds of stuff the Japs had grabbed from the people on Cebu. I saw those and thought God, I hope we don't have to clean out those suckers." Chuck Walker examined a cave filled with generators, wheel barrows, and sewing machines. Another cave held a "nearly new" American automobile. Huge piles of Japanese occupation money were everywhere, now all but useless.

Over the next several days, patrols spread out to find and kill any Japanese that could be found. General Arnold's divisional report estimated that his men had killed almost 9000 enemy, but the official history casts doubt on that, noting that thousands more surrendered on the island in late August 1945. The division's own losses amounted to 410 killed and about 1700 wounded. This was more than could be replaced, and the division faced another fight on the island of Negros.[11]

The 3rd Battalion was taking care of its business on Bohol. It landed on April 11, meeting no opposition. There were only about 300 Japanese on Bohol. At first no one could find them. Paul Guzie recalled seeing just a few in the first week. "While moving inland we came upon several Japs about 700 yards away bathing in the ocean, did some firing, but because of the distance I don't think anyone was injured." Marching into a barrio, they set up a roadblock and spoke to a few inhabitants. Most of the civilians were hiding from the Japanese. But word got out quickly about the Americans' arrival. Within hours men, women, and children were pouring into the villages. A big "liberation celebration" got under way. Guzie missed most of the "drinks, games, and Filipino music for dancing" because he had guard duty. Local alcohol was called "Tuba juice," made by boring a hole in a coconut, filling the hollow with sugar and then leaving it to ferment. "It was like 100% alcohol." The festive atmosphere was soured when Guzie saw a Filipino guerilla at the barrio's makeshift hospital, "cut nearly in half by machine gun bullets." He died the next day. Guzie also saw starving small children who died even as the soldiers tried to feed them warm milk.[12]

Around April 15 a 3rd Battalion patrol located most of the Japanese entrenched in a series of hills "about eight miles inland from the center of Bohol's south coast." The festivals were over. K Company moved into positions with L Company around the enemy's main defense line. The Japanese were alert and the fighting got vigorous. Guzie and his friend, A.K. Hall, found a pillbox and lobbed grenades at it. Hall then took it under fire from a standing position, yelling "they can't hit me," while other men got close and destroyed the pillbox. Hall received a medal for this.

The enemy counterattacked that night but was driven back. Company I then flanked the enemy line and forced them into the open. Their position broken, the Japanese scattered in small groups. Patrols ferreted out many over the next few days. Guzie and a few comrades came on a small group of Japanese with a light machine gun outside a small cave. "Two of us got above them and threw grenades but they would dive back into the cave." Going back to a small Filipino village, Guzie and his pal got "some glass jars." "We could drop a grenade into the jar and the glass held the safety lever in place. When we tossed it down the glass broke, the grenade rolled into the cave. That got them." Guzie was given a Bronze Star for this. By April 25, the 3rd Battalion had reduced the Japanese to a handful of fugitives. Their own loses were seven killed and fourteen wounded. The three companies were back on Cebu by the end of the month.[13]

Back on Cebu, the Americal Division made its preparations for landing on Negros. By 1945, the American Army had troops in action over much of the globe. Its manpower shortages were becoming serious, forcing commanders to strip service units for infantry replacements. In the Philippines they would have to keep relying on very weary veterans who had been fighting since 1942. Until divisions from the European theater arrived, there were no fresh units or experienced officers to take their place.[14]

Chuck Walker was awarded a cluster to his Silver Star and given a new job. The 2nd Battalion's executive officer had been transferred to a new assignment. Mahoney told Walker to report to battalion HQ as the new exec. Battalion commander Mac was anything but happy. "Looking at me as I walked in, he said grimly, 'I guess we'll have to learn to get along together.'" Walker retorted, "'The problem is yours; I've done my best.'" Years later Ross reflected that "Chuck Walker taught me how to be a good officer. He had brilliant tactics, up front leadership, and physical and moral courage. He had the ability to have been an outstanding division commander. But he challenged the officers in charge and they resented it."[15]

The 164th was the most experienced regiment in the Army's Pacific forces. Cebu/Bohol was its fourth campaign and several experienced men were already gone. Al Wiest had left Bougainville for the U.S. to teach updated machine gun tactics. "The colonel at my first post in the U.S. was shocked I hadn't gone to the Infantry School at Fort Benning," Wiest remembered. So off he went to Benning. "They needed my instruction badly. I got on the machine gun committee and before long those of us with combat experience got rid of all that indirect fire stuff." John Tuff was offered a chance to go home. But he decided to stay on for the expected invasion of Japan. He had turned thirty in 1944, an old man by Pacific warfare standards. Illness sent still others to the States. "I had healthy parents," Doug Burtell reflected forty years later, "and was healthy and strong. I did have malaria and dengue fever, but didn't have jaundice or dysentery like a lot of the fellows. Some of the guys that did have that had been 250 pounds and were down to 150 pounds. They had to fly them home." Burtell himself thought his next battle would be in Mindanao or China or Formosa.[16]

As replacements came in, old vets moved to less dangerous duties. As a sergeant, Joe Castagneto had his cap shot off in one battle and later was wounded by an American mortar round that killed another man and fragmented his legs. "My knees burned like heck. I thought I'd shoot somebody. But Gene [Scolavino] who bandaged my legs, took my rifle." He had a reputation for taking good care of his men. On Bougainville, he got cans of tomatoes and a large box of spaghetti from home and cooked it on a mountain stove. "The aroma was in the air and we had visitors from the company come around with their mess gear … but there wasn't enough to feed these hungry dogs." He made spaghetti again on Cebu when a cook stove exploded and destroyed the planned meal. After one of the cooks was burned in another stove mishap, Castagneto was "detailed as a cook until we returned to the Good Old U.S.A."[17]

John Paulson thought "all combat experience is the same but never in the same pattern. When you're responsible for other people's lives, you have to make a decision, that's the thing that's bad about combat. You do the best you can because you don't want to get anybody hurt if you can possibly prevent it. You get caught up in the fight with them. There is no such thing as a nice battle I'll tell you." By the time Sergeant Paulson reached Cebu, he thought of war as "a sad dull routine, it was same thing, the shooting and the killing." Then on Leyte he was "posted on a ridge. I was looking down a road and saw about ten Filipino women pulling a Jap sailor along." They told Paulson they were going to kill him. Paulson said 'I can't let you do that.' One woman cried out that they had 'lost a lot of people' at the hands of the Japanese occupiers. He replied, 'So have we. We've lost a lot of Americans getting here, so you'd be free. You let me take him.' He finally ended the matter by getting two riflemen to take the sailor, saying 'If I let you kill him, then we're no better than they are.' He took the prisoner to the F Company compound and got him some food. "He looked exhausted and starving. And the cook griped about it, complaining 'we chase these guys all over the Pacific and now I've got to feed them.'" Looking at him later inside the prisoner's wire, Paulson wondered "if God had kept me alive all this time so I could really forgive my enemies and save this man." By then there were only about ten men left from his company as it was in Carrington before the war. The rest were in other units, recovering from wounds or dead.[18]

Filipino civilians complicated the American's mission. S2 officers of the 164th made detailed lists of the looted goods they found in caves, but no one knew how to redistribute it. Soldiers gave Filipino women sewing machines they took from the Japanese loot – and it wasn't always traded for the repair of a torn shirt or clean laundry. By the end of April, medical officers were noting a "significant" increase in venereal diseases among the ranks in all the army and navy units in the Philippines. In addition to using some of the booty for sexual transactions, some sold loot on the black market. Hungry people are desperate and some Filipinos stole from the soldiers. Walker recorded one incident in which a man trying to steal a barracks bag was shot by soldiers and his body left in the street as a warning. John Paulson's experience with the angry women underscored the issue of revenge, both on the Japanese and also on one another. Soldiers had to intervene in many cases where a group of civilians were about to kill a neighbor. Was he a collaborator or was this a personal score? When an army came to liberate, more often than not it stayed on to administer. Maintaining order took time away from beating the Japanese and winning the war.[19]

Southern Negros was the next assignment. This part of the island was a rugged patch of rocky hills that rose to as high as 6000 feet. Only near the coast was there any significant open ground. It was a defenders paradise and the Americal Division intelligence estimated the enemy force on it at about a thousand to fourteen hundred men. The divisional headquarters ordered the 164th to assume the enemy would contest the landing. There was still a shortage of landing craft but the engineers went to work and decided that the craft could be overloaded without endangering the occupants "in normal running seas" (one wonders what the occupants would have thought about that, if they knew it).

Taking this information, Mahoney's staff drew up a schedule that would place most of the 1st and 2nd Battalions, with supporting artillery, engineer, and medical support, on the beach in five waves. The orders noted that in order to keep the supply needs within the restricted carrying capacity of the craft, "two days [of] K rations will be carried by all troops, under no conditions will these be used as voyage rations and later will be consumed only as directed by Company COs." The regimental supply officer was to keep careful control of all subsequent needs, and all vehicles were to carry only three-quarters of the normal gas capacity. Other necessities were also carefully limited.[20]

The force embarked a half-hour before midnight on the 25th of April. The landing craft were screened by three destroyers as they made the run to the shore. The seas cooperated by being relatively calm, a boost to the morale of the landlubbers crammed into the rolling craft. One of the destroyers lobbed a few shells onto the beach, while the first wave of 1st Battalion headed for the surf, wondering what waited for them. But Colonel Satoshi Oie, the commander on Negros, had chosen to make his stand in the hills. The 164th went over the beach standing up.[21]

The other waves followed quickly, although the tight timetable was altered to allow removal of some "mostly improvised mines." A second echelon was expected in two days. Charles Walker organized vehicles for the 2nd Battalion while riflemen double-timed toward the town. In Dumaguette there was no opposition. The Japanese had pulled out as soon as the American ships appeared, burning supplies and parts of the town. The oncoming GIs killed a few as they ran. Many of the Japanese, observers noted, were "not in complete uniform." McConnell organized part of Company E to push west and try and cut off the fleeing enemy. "We had verbal orders to shoot first and ask questions later," rifleman Zane Jacobs noted. Jacobs and his platoon moved west from Dumagette.[22]

Zane Jacobs had been drafted in August 1944 in Oklahoma City and sent with many others to Fort Sam Houston, Texas. "At the induction we were all interviewed by the Army, Navy and Marine Corps. I asked to go into the Navy but of 78 men that day only two were allowed into the Navy. The rest were into the Army." He went with a host of others to Fort Hood for infantry training, in "over size or balloon companies, there were 250 men in my company." Every third man was culled after about twelve weeks and sent off. Jacobs heard they went across the Atlantic and into the Battle of the Bulge.

After completing his training, Jacobs got a short leave to see his parents and then was off to California and onto what he was told was "a converted cattle ship" for as quick a dash to the Philippines as the old tub could manage. The usual cold chow, queasy stomachs, lifeboat drills, and salt water showers became the daily routine. The ship called at Hollandia in New Guinea to pick up a group of veteran infantrymen who had recovered from wounds and illnesses. Jacobs quickly noticed that "they kept to themselves for the most part."

They arrived at Tacloban and went into a "repple depple." Jacobs and others received brand new M1s. "It took hot soapy water to get the cosmoline off all the parts of the rifle." Jacobs turned nineteen a few days later, on April 12. After briefings on tropical diseases and the Atabrine regimen, he. was added to the roster of Company E of the 164th. "All the companies were undermanned. I was assigned to a platoon with 18 or 19 men in it and my squad had five men. Jack Kachel was the platoon sergeant. He was from Arizona and a regular army man. He made me second scout of the squad. Less than a week later we boarded LCIs and headed for Negros." Now, having reached a point west of Dumaguette, Jacobs's squad "set up a perimeter across the road that led into the hills. About midnight three trucks loaded with Jap soldiers came through the ambush. The city side of the perimeter waited until one or two trucks passed and then opened with their .30 caliber machine guns. We opened up when the trucks passed the west side of the perimeter.. We stopped all three trucks and next morning we counted 18 to 20 dead Japs. We took no casualties in that skirmish." [23]

The mountain bastion that Oie organized was in very rough terrain. Scout patrols went out to sniff out Oie's positions in the mountains. Two Japanese "infiltrators" were caught and killed as they tried to sneak up to a company command post. They were killed after getting off just one shot from an old Enfield rifle. The GIs found TNT on the bodies. This was the first encounter the regiment had with a "suicide squad" on Negros, organized by Oie to harass the American soldiers. Many more such bands armed with old weapons and explosives made attacks in the days ahead. When cornered they often killed themselves by holding grenades up to their heads.[24]

On the 28th the second echelon of troops landed at the beaches, carrying mostly artillery and more supplies, plus some tanks. E Company, following a lead given to them by Filipino guerillas, headed up a long escarpment into the hills west of Dumaguette. They found abandoned trucks that ran on alcohol and some tossed off Japanese equipment. On the 30th they made contact with the enemy. Zane Jacobs's squad was moving up "near a knob of the ridge" and came under fire from an enemy machine gun.

A vet told Jacobs it was a Nambu. "Lazono was point scout. He dove across an opening in the underbrush and a bullet cut across his forehead, temporarily blinding him. We never saw him again after we got off the hill and he went to an aid station. A tall Texan named Hale was our BAR man, he stepped out and opened up on the Nambu, stopping it a couple of seconds while they replaced the dead gunner. The new gunner cut loose a few rounds and hit one of our men in the ankle. Hale took out the second gunner and then he carried the man with the shattered ankle back."

Another Japanese took over the machine gun and pinned down Jacobs' squad for almost another twenty minutes. A GI gunner about half a mile downhill made it worse by pouring counter fire up at the enemy nest, a number of rounds falling short around Jacobs's squad. This shook a man named Utah Christopher, "a seasoned veteran." Caught between the enemy and friendly machine guns, he tried to crawl back. Jacobs was in a lower spot, the enemy gunfire passing "about three to six inches above me. I told Christopher to calm down and we might find some way to get through the brush without being seen and maybe wipe out the gun. But 'Chris' tried to crawl back in the grass and the movement of the grass "gave his position away." A burst tore through the grass. "I heard the bullet hit Chris and heard him die. It didn't take long." Then Jack Kachel came up on the flank of the Nambu and hosed it with well placed fire, killing the enemy crew. The squad's survivors pulled back, but Kachel had Jacobs help him carry back Christopher's body. "I was not anxious to go back but I was the only one who knew where Chris was located. He had a .25

round through his chest and a gaping hole big enough to put a fist through." Kachel told Jacobs that "we never leave dead or wounded."[25]

Young soldiers grew up fast in these battles. Crescencio Cruz was just twenty when he landed on Negros. Raised in Arizona, Cruz was the son of a gardener. After being drafted and trained in 1943, he joined Company F of the 164th on Bougainville. "I was scared all the time," he admitted, and went through several exchanges of fire. "One time Leonard Cortez and I were leading the patrol and a Japanese machine gun opened fire on us from the jungle. Cortez was hit in the back but didn't know that. We crawled back and the patrol withdrew. This guy asked if I would carry his pack and I told him I was too tired. A bit later I realized he was bleeding. I said 'why didn't you say you were wounded, dummy,' and he said he didn't want to be a bother!"

Cruz was one of a number of Hispanics to join the North Dakota regiment as replacements. "They took six of us and put two each in the three platoons. It was kind of funny with role calls, you know – 'Castro, Clauson, Cruz, Espinoza, Swenson, Torgerson' – it sounded really strange." But his comrades appreciated his language ability. "Because I spoke Spanish I could talk to the civilians which helped because we had to figure out who gave us good information and who was for the Japs." Cruz fought in the battles on Leyte and Cebu where several of his friends were killed or wounded. He carried two seriously wounded men back from Babag Ridge. "William was bleeding from his face and Albert Dietl had shrapnel in his body. I carried them to the aid station."[26]

On Negros, Cruz went into the hills with his platoon. The unit soon "got hit with some mortar fire. A guy in our unit by the name of Napthali Carter got hit when a shell exploded in front of us. His mother was Hispanic. He got hit in the face and I helped when they started to carry him down the hill, but he died in my arms as I carried him. I see that in my mind all the time. I know a lot of men have gone back over there to see those places again, but I have no desire to do that. Sleeping in those jungles, the fighting, the guys killed, I have no fond memories of the Pacific."[27]

Ben Rosalez was another man who came into the ranks as a replacement. "I was eighteen when I got to Leyte and was assigned to a rifle squad. Things got a little mixed up in the C Company tent camp where I saw a cot with a card that said 'Ben Rosalez.' I thought that was pretty nice. Then this guy comes in and says 'why are you in my bed?' and it turned out he was also Ben Rosalez, from Arizona, and had been there since Bougainville. They got us mixed up a few times."[28]

Negros was Rosalez's first action. "We had a sergeant in the platoon who was about to go home but as we went up to the Jap lines, he was killed when he went behind a tree to relieve himself and a shell came down. It might have been one of ours that fell short but it didn't matter." His first combat came when his platoon tried to push the Japanese out of a position that was covered by a "sheer rock wall." "We couldn't really climb that so we used grenades and so did they. I shot one man who was up to throw a grenade at us. It went off as he fell and killed some others. I threw several grenades up and over the rocks, counting five each time so the grenade would go off right over the wall. I heard their screams as they died." On another occasion the Japanese counterattacked. "It looked like the banzai attack you heard about from earlier in the war, them screaming and us firing and driving them back." An enemy grenade exploded near Ben's foxhole and fragments cut his cheek. "That was one time I got hit and another time I had my back burned when one of our phosphorous grenades hit a tree." He treated the injuries himself and never asked for a Purple Heart.

Company C's position, in the mountains west of Dumaguette, hindered resupply. Rosalez watched "light airplanes drop food and water to us in shell cases, but the Japs got some of this." The fighting was brutal and at night it got chilly. Men had to huddle together to keep warm. "The worst thing that happened was when the Japs captured one of our guys during an attack. "They tortured him to death above us, a man of theirs yelling in English 'this is what we do to American soldiers.' We could hear his screams and a day or so later we moved up and found his body tied to a tree. He had been bayoneted many times. After that our men didn't care to take prisoners."[29]

April moved into May and the two battalions scoured away at the Japanese defenses. Down on the coast, small groups of Japanese used boats to infiltrate Dumaguette and wreak havoc. One group of two dozen enemy soldiers landed from three sailboats and tried to attack a depot. Two companies from another regiment were sent to track them down. A day later, Filipino guerillas patrolling the coast in a Navy LCM spotted "3 bancas," (a Filipino craft) carrying some sixty Japanese soldiers. They shot up the boats killing most of the men.

"Of the five enemy known to have reached land, two were killed and three escaped to the hills." The report of this incident concluded "one Jap captured, later killed, stated they came from Bohol," apparently to reinforce their comrades in the mountains. Small groups of Oie's men tried to slip through GI positions. These suicide squads were wiped out when they were discovered by patrols or guards.[30]

This kind of warfare strained the men's nerves. Everyone concluded that if the enemy was going to fight like this for a mountain ridge that was meaningless, then Japan would never surrender. Fatalism permeated the enlisted men and officers clashed as they tried to find another way to break the Japanese resistance. Engineers built some airstrips in the foothills so that light aircraft could direct artillery fire. Riflemen in the hills would watch as the planes, far below their lines, took off and slowly climbed to higher altitude. Bulldozers were brought up to build a road that would allow tanks to support the infantry. Chuck Walker watched a "cat dozer" at work. It lost traction and slid on the "thin carpet of topsoil," almost going over a cliff. Walker "hooked three trucks with winch cables to hold the cat until we could get it started." McConnell came to see the situation and told Walker "that if I lost those trucks my ass was mud." The dozer was winched out and went back to work.[31]

A much worse fracas came soon after that raised tensions further. Colonel Mahoney, frustrated by "the strong resistance of some of the Japanese defenses," ordered the 1st Battalion to pull out of its existing positions in the hills and move around the heights so as to attack from the north while the 2nd Battalion struck from the east. It was a bold proposal but success would rest on speed and precise timing. This set off a quarrel between Mahoney and his battalion commanders. John Gossett thought the whole concept was flawed, saying that it would be too difficult to supply or support his troops across the distance involved, and that his men would give up ground that they would have to fight for again once the flank move failed. Gossett added some personal remarks that questioned Mahoney's abilities. McConnell was more circumspect but he too doubted that the maneuver could work with the number men that the 164th had at hand. Mahoney then gave the men a direct order and the 1st Battalion fell back to wait for trucks to haul them over the narrow coastal road to the north side of the mountains.[32]

    The flank move turned into a tragicomedy. With only poor maps of the area to go by, the trucks got lost and arrived late at the drop off points. The companies then had to march nine hours to reach a "mountain village" that was to be their jump-off point, but, again due to faulty maps, the men could not find the village. Worse, the Japanese, holding the higher ground, saw the American movement. While the GIs took a long arc to get to their new positions, Oie shifted his men across the much shorter chord within the arc and cut the 1st Battalion's supply line. The attack was stalled before it began. Another group of Japanese moved in to retake the ground the 1st Battalion had given up, as Gossett had predicted.

    This however did weaken the enemy's defenses in front of the 2nd Battalion enough to allow G and E men to move forward and gain some substantial ground.

Shifting his own men back toward the east, Gossett was able to link in with McConnell's forces and reconnect into the united front. The supply lines were reopened and the 164th was now facing Oie's main defenses. Oie had no access to supplies now and the Americans hoped he would choose to run rather than take "a pounding by artillery and bombs while starving to death." Hope that the enemy might break arose the ranks. In F Company, the daily log for May 2 stated: "rumors have it that the operation is almost over . . . only enemy activity is occasional sniping of stray Nips. Also got mail, which is a life saver out here."[33]

The maneuver had not worked as intended, but it drew Oie out enough to make him vulnerable. In the days ahead his men were unable hold back the attacks that now began across the whole of his front. But the disagreements between Mahoney and his commanders had done damage at the headquarters. A few days later Mahoney sacked his S3 and the exec, reassigning them to the 2nd Battalion. Soon after that, Mahoney ordered John Gossett to attack a hill with what he regarded as an insufficient force. He refused and Mahoney relieved him. Chuck Walker and Veon McConnell reached a rare accord and wrote a complaint concerning Mahoney's flawed maneuver. But the 164th was advancing by the time the divisional staff looked it over, and the complaint was quashed. Mahoney then put Walker in temporary command of the 1st Battalion.[34]

Oie continued to hold out. Zane Jacobs watched F Company as it engaged in a struggle to take another Japanese machine gun nest. As Jacobs and a group of Filipino supply haulers he was escorting got nearer to the action, "Fox company opened up to return fire from up the hill above them. We stopped the Filipino bearers behind a small ridge till the fire died down." He looked around for a vantage point to "get a better picture of what was happening above us" and while doing this found a dead Japanese soldier. "He must have taken a direct artillery hit as about all that was left was his torso." One of the man's arms "with hand attached" was hanging in a tree above the spot. Jacobs found a Japanese battle flag on the man – "he must have been a platoon leader." The flag was "covered with blood and brains but I took it as a souvenir" as well as a Colt pistol he also saw on the ground. Jacobs wanted the flag for a gift. "Tom Purvis had been with the 164th since its Guadalcanal Campaign and had never gotten a memento of his fighting except malaria. He was going home to Baird, Nebraska, so I gave him the Battle Flag."[35]

Jacobs did not expect to go home soon, if he got to go home at all.

(Endnotes)

[1] Cooper and Smith, *Citizens as Soldiers*, pp. 315-16; Cronin, *Under the Southern Cross*, pp. 295-96. The "After Action Report, Americal Division, V2 Operation," contains much useful information for each of the Bohol, Cebu and Negros operations. A copy of it is in 164th Infantry Association Records.

[2] Walker, *Combat Officer*, p. 188.

[3] S2 Journal entry for April 12-13, 1945, and Robert Ross Smith, *Triumph in the Philippines*, pp. 616. Cooper and Smith, p. 317, mistakenly refer to this part of Cebu as "Bagbag Ridge." Cronin, pp. 295-97, and Philippine sources make clear that Mount Babag is the dominate feature of the "Babag Mountain Range."

[4] Robert Ross Smith, pp. 616, cites in addition to Arnold's postwar commentary the opinion of the Japanese Chief of Staff of the 102nd Division, who viewed the 164th assault as poorly conceived, "had started too late and had been too weak" to gain a breakthrough.

[5] Combat summary prepared by Major Wendell Wichmann, for April 13-14, 1945, attached to S2 Journal, and April 14 entry in the S3 "Operations Report: Cebu/Negros," p. 1-2, both in 164th Infantry Association Records.

[6] Wichman summary for April 13-14, 1945; Walker, pp. 188-90; Cronin, pp. 297-98.

[7] Cronin, pp. 295-96; Cooper and Smith, p. 317; S2 Journal entries for April 14, 1945, and combat summery prepared by Major Wendell Wichmann, for April 14-15, 1945, attached to S2 Journal.

[8] Wichmann interview with Shoptaugh; S2 entries for April 14-16, 1945; April 16, 1945 entry in the S3 "Operations Report: Cebu/Negros," p. 2, 164th Infantry Association Records.

[9] "Company F, 164 Infantry: Log of This Unit's Activities from 11 April 1945 to 20 April 1945 Inclusive," entries for April 11 to 16, 1945. This log was preserved by Captain William Jordan, F Company commander after May 5. Jordan's son, John provided a copy to Shoptaugh. Some punctuation is added for clarity.

[10] S3 "Operations Report: Cebu/Negros," p. 2-3, 164th Infantry Association Records; Walker, p. 192-93; Cronin, pp. 298-99.

[11] Castenegto/Scalavino interview with Shoptaugh; Walker, pp. 193-95; Robert Ross Smith, pp. 616-17. Smith accepts that some of the Japanese who surrendered on Cebu in August were "late escapees from Leyte," but those surrendered on Leyte also exceeded the estimates made by the 8th Army.

[12] Guzie, "From New Caledonia to Japan, 1943-1946," together with Shoptaugh interview of Guzie, July 7, 2007.

[13] Guzie interview with Shoptaugh; Cooper and Smith, pp. 318-19 summarizes the Bohol actions, stressing that the Japanese commander "had little rain forest or jungle to conceal troops with."

14 Maurice Matloff, *Strategic Planning for Coalition Warfare, 1941-1942* (Washington DC: Center for Military History, 1953), pp. 113-19, 178-82. Jay Luvaas and John F. Shortal's essay "Robert L. Eichelberger: MacArthur's Fireman," credits Eichelberger for making the most of limited resources, in *We Shall Return!: MacArthur's Commanders and the Defeat of Japan, 1942-1945* (Lexington: University Press of Kentucky, 2004 reprint edition.) pp. 155-176.

[15] Walker, p. 196; Ross quotes from Shoptaugh telephone interview with Charles Ross, September 2008 and Ross, "Tribute to Charles H. Walker," *164th Infantry News,* December 1997. Ross himself made the army his career and commanded combat forces again in Korea and Vietnam, including airborne ranger and Special Forces.

[16] Albert Wiest, Edith Tuff and Douglas Burtell interviews with Shoptaugh; Burtell, written memoir, 2003, 164th Infantry Papers.

[17] Joseph Castagneto, "Spaghetti ala Bougainville," and "Swiss Steak that Blasted Off," *164h Infantry News,* July 2006 and July 2007 issues; Castagneto/Scalavino interview with Shoptaugh.

[18] John Paulson, interview with Shoptaugh, October 20, 2007; "Company F Reduced in Number Since Leaving Carrington in 1941," *Foster County Independent,* May 20, 1944.

[19] Lists of cave loot in S2 Journal entry for April 19, 1945; Walker, p. 195 (killing of thief); Cronin, pp. 35-38 for general administrative challenges.

[20] All the planning documents, including debarkation schedule, landing diagram, and invasion field orders, are attached to the S3 "Operations Report, Cebu/Negros," 164th Infantry Association Records.

[21] Cronin, pp. 315-19; Cooper and Smith, p. 320.

[22] S2 Journal entries for April 25, 1945 (quoted); Walker, pp. 196-98; Jacobs e-mail to Shoptaugh, November 3, 2009.

[23] Here and below, Zane Jacobs e-mails to Shoptaugh, March 5, 2007 and November 3, 2009.

[24] S2 Journal entry for April 27, 1945.

[25] Zane Jacobs e-mail to Shoptaugh, March 5, 2007.

[26] Here and below, Crescencio Cruz, telephone interview with September 28, 2007.

[27] Cresencio Cruz, telephone interview with Shoptaugh. Sixty years after the war Cruz learned that the son of James Simpson, a Company F man killed on Leyte, wanted details about his father's service and contacted him. See *164th Infantry News*, March 2006.

[28] Here and below, Ben J. Rosalez, telephone interview with Shoptaugh, September 5, 2007.

[29] Rosalez, telephone interview with Shoptaugh.

[30] S2 Journal entries for May 1-2, 1945.

[31] Walker, p. 200, with Walker e-mail to Shoptaugh, June 8, 2009.

[32] Cooper and Smith, p. 322; S3 "Operations Report, Cebu/Negros," entries for April 29-30, 1945; Walker, p. 209. Walker states that McConnell suggested that the regimental exec, Bill Considine, cancel the order because Mahoney was drunk. Rumors of Mahoney's drinking had been rife in the regiment, but the maneuver could not have been organized without cooperation of the regimental S3. What bearing this incident had on Mahoney relieving Considine and his S3 a few days later is unknown.

[33] Cooper and Smith, p. 322 (for Oie's choices); May 1-3, 1945 entries in S3 "Operations Report, Cebu/Negros," report (for 2nd Battalion progress); "Company F, 164 Infantry Log," entry for 2 May 1945.

[34] Shoptaugh interview with Wendell Wichmann (who tried to mediate the argument between Gossett and Mahoney); Walker, pp. 210-12. In a September 2008 interview with Blake Kerbaugh, Walker stated that Gossett "refused to have a bunch of guys commit suicide for the Old Man" (Mahoney). John Gossett was transferred to the U.S. at the end of the Negros campaign. He left the Army in September 1945 and entered into a very successful hardware business in California after the war. Almost every man who served with him regarded him as one of the best of the company and battalion commanders.

[35] Zane Jacobs e-mail to Shoptaugh, March 5, 2007.

Chapter 15: "The Greatest War in History" Negros-Japan: June-November 1945

"**November 29, 1944, Australia**, Dear Love, I am writing this from Townsville, Australia. We arrived by flying boat at noon . . . Today for the first time in many months, I looked upon civilians. It was quite a sight seeing family life functioning in routine manner – father on his way to work, the kids playing in the street, mother doing her day's shopping, etc. . . . For a man straight from the front this land is really a paradise." Milt Shedd

"Somewhere in the Philippines, April 22, 1945: Dear Mary, I haven't had a good chance to write a letter in a couple of days due to the military situation . . . I got a card with a dollar in it from the VFW. I don't need the money at this time but it was a nice new dollar so I kept it for a reminder of the 'old country.' Love to all." George Dingledy.[1]

"Home alive in '45" was the slogan used by the GIs in Europe. It meant nothing in the Pacific. General George Marshall believed the Pacific war could last until 1947. George Dingledy doubted that and remembered that he and his friends were using the phrase "the Golden Gate in '48."[2]

The Joint Chiefs in Washington disagreed on how Japan could be forced to sue for peace. The Army Air Force commanders believed they could bomb Japan into submission by burning the enemy cities with fire bombs. Admiral King, the head of the Navy, thought his ships could blockade the Japanese home islands and starve them into surrender. The Army planned a "Japan Campaign" that was to be carried out in two phases: the first, an invasion of the southern island of Kyushu, using a task force even larger than that employed at the D-Day landings in France. An estimated forty-two aircraft carriers, twenty-four battleships, and four hundred destroyers and destroyer escorts would accompany thousands of transports carrying fourteen U.S. divisions for the initial landings. This was scheduled for October 1945. Then, if Kyushu were subdued in three to six months, the main island of Honshu could be invaded with no less than twenty-five divisions. After Germany's defeat, at least fifteen divisions were to move from Europe to reinforce the forces already in the Pacific. Britain and Australia would contribute additional forces. The Japanese were expected to resist by using their armies, thousands of suicide aircraft, suicide boats, and even civilians trained to fight as suicide bombers.[3]

Planners disagreed on the number of Allied soldiers they expected to fall in this campaign. A study by the Joint Chiefs projected 1.2 million total casualties, with some 300,000 killed. MacArthur's staff reduced this estimate by about half – provided MacArthur had full control of all the forces, they stressed. A separate study conducted by staff for the Secretary of War was more pessimistic: at least two million Allied casualties, with 4-800,000 killed. It also estimated Japanese dead, including civilians, at five to ten million. Other projections gave varying estimates. Obviously, no one could be certain of anything but that it would be a very bloody and protracted fight.[4]

It was small wonder then that the Army Air Force employed every weapon at their disposal to reduce Japan's industry and demoralize the Japanese population. After conventional bombs failed to produce enough damage,

B-29 bombers used incendiary bombs. In February they incinerated a square mile of Tokyo. Additional raids torched more of Tokyo and wrecked several other cities. But even though civilians were killed in increasingly large numbers, the Japanese still showed no signs of giving in. After Roosevelt's death, President Truman engaged in negotiations to bring Russia into the war against Japan. Even though he did not trust the Soviet Union, nothing could be ignored that might reduce the casualties of an already war weary nation. Truman learned on July 16 that the first prototype of an atomic bomb had been successfully detonated in New Mexico. He did not hesitate in ordering the military to use it if the Japanese refused to give up.[5]

Out in the Philippines men knew only a little about the B-29 raids, were ignorant of the negotiations with Russia, and had no inkling about the revolutionary atomic bomb. They had heard rumors about the high casualty projections for an invasion. What they knew best was that, in defiance of all military logic, the Japanese were still willing to die in defense of meaningless ground. Home alive in '45? They found the idea to be laughable.

By third week of April, the 164th had reached the heart of the Negros defense line. It was in a series of caves and interlocking strong points in a high ridge that was part of the face of Mount Cuernos de Negros (horns of Negros). Because of the daily miserable rains in April and May, the GIs called it "rain forest ridge" (see Map 10). The highest point of the mountain lay between the infantry that was assaulting the enemy and the artillery that could have supported them. Only a perfect shot could clear the crest and fall on the enemy. Any error in trajectory was either wasted on the rocks worthlessly or fell on the American footsloggers. This exacerbated bitterness over the decision to shift the line of attack from east to north. Here the "five Fs" were reduced to two – fighting and finishing. They had to dig out the Japanese with close-quarter assaults.

PFC Bob Love was a machine gunner attached to B Company for the Negros fighting. "We fought the Japanese from one hill to the next, each one getting harder to take. When rats get cornered they tend to fight harder. They had dug in deep [holes] and caves on every hill with plenty of supplies. There was lots of blood and bandages in each cave [taken] so we must have given them a lot of hurt." Once again, the enemy had disguised their positions well. The GIs used flamethrowers when they could. "One night we dug in, in case of a counterattack. Our artillery observer heard some shells coming in, and eight rounds of 90mm Jap mortar shells came in, walking up a line right towards us. The last one hit about 200 feet in from the HQ tent. We were lucky."[6]

Swede Clauson was also still with B Company, providing support fire for attacks. After surviving Guadalcanal, Bougainville and Leyte, he was convinced that his odds had to run out soon. They were now clawing away at another knob filled with tough defenders. "We could even hear them talking but couldn't get at them. The hill was filled with rocks and boulders, it had no military significance whatsoever, and they could have held us off with three or four guys. Three times we had tried to take it and each time we were forced to turn back from withering fire." Someone at the company or battalion level suggested they were not trying hard enough. The word 'yellow' was mentioned. Now the squad was grimly determined to fight it out to a finish, whatever the outcome. "Those of us who had been fighting for close to three years had seen a lot of our buddies go down. It was obviously our time to go."

The night before the attack he sat talking with his squad leader, acting Corporal Joe Acosta. "He had been a sergeant a couple of times but had been busted for reasons best not disclosed." Joe noted that the war was ending in Europe: "at least we've lived long enough to know we've won. Remember that time you got drunk in Fiji and chased one of the girls up a tree and cried because you didn't know how to climb a tree?" "Yup," Swede replied, "how could I forget, that story has circulated all over." "Tomorrow when we storm that hill, I want you to climb that tree." "Okay," Swede said, "we go out in style, no regrets, no crying." He and Acosta shook hands. "You've been a hell of a soldier." "Yup, and so have you." They did not say much the next morning and ate only a bit; "nothing tasted good." Then an officer came around to say the attack had been called off. Aircraft would paste the knob first and then it would be attacked with flamethrowers. "Joe told me he wanted me to lead the detail back." A few days another platoon had several men killed in taking the knob. Clauson and Acosta did not see combat again.[7]

Over in Company F, men watched in horror as G Company assaulted a very stubborn group entrenched on another knob. "After the heaviest artillery concentrations of this operation," G's platoons went up a rock-strewn slope "too steep and narrow for one [entire] co. to push at a time." The men could only approach in squads. "After six hours of rough uphill fighting, they almost reached the top and were forced to withdraw when the Japanese "opened with mortars and machine guns." The entry concluded, "G Co. has their hill to take . . . and the next one is [ours]." G Company took its position after two more attacks, then moved on to blast out another group of defenders in a shack further up the mountain. Discovering a cave under the shack, one of their men took a bazooka and fired a round down the shaft. The "cave was apparently an ammo dump as there was a violent explosion followed by continued minor explosions."[8]

F Company then began its own assault. It did not start on a promising note. Captain Jordan, now the company commander, could not understand how the Japanese withstood "our artillery bombardments." He thought the shock alone had to incapacitate some of those men in the caves. Perhaps so, but that night an enemy suicide squad launched a counterattack. One of Jordan's men was hit by "small arms fire (friendly) and died a few hours later." A group of GIs prepared to attack the next morning. They were bitter that "many of our best men [were being] either killed or wounded," especially since "rumor circulates that Americal Division will soon return to U.S." Perhaps in was anger that spurred them on the next morning. After Marine Corsairs dropped bombs and strafed the targets, the men went up. One platoon seized a knob half way up the slope while another platoon followed and pushed on.

Casualties so far were light. The next morning several men were wounded, but they took the height. "Everyone was surprised at us taking the hill as quickly as we did." Better yet, the height gave them a chance to pour fire onto another position the Japanese refused to give up. "For once we were looking down their throats and gave them hell, killing quite a few." The enemy fled and then attacked again that night, "Seven times with grenades, TNT, rifles and machine guns, but we repelled every attack without suffering any casualties. Nip casualties were heavy but they drug their dead away before morning." Fox Company repelled additional attacks for the next three days, then expanded their hold on the hill and burned out more caves with flamethrowers and grenades.[9]

Finally Colonel Oie's defenses were buckling. The artillery fire and aircraft strikes did their part to break them but it was the courage and grit of the infantry that blasted Oie's men out in the end. Before a ring closed on all of his force, Oie withdrew small groups, two or three dozen at a time, while the rest of his men were sacrificed. When Company B advanced on the 17th, they noted that even when in well entrenched spots, the Japanese fire was at times "very inaccurate and erratic." This was probably because Oie used wounded and sick men to act as his rear guard and because by now the Japanese were very short on food. Oie's officers still sent out 'forlorn hope' suicide squads, one of which was wiped out while trying to destroy a bridge outside Dumaguette. On the 18th a patrol captured a man who told them that Oie had abandoned his headquarters, which he said was "about two hours from 1st [battalion] observation post." Bombs had demolished it, he said, and added that the Japanese "had very little food left." The 164th spent the rest of the month finishing the job in those mountains.[10]

In the last days, the GIs were finding just one or two Japanese at time, mostly wounded, sick or starving. Several of these men killed themselves when they saw the Americans. One of the suicides wounded Doug Burtell. "Larry McCarten and I were returning from a patrol and saw the kitchen truck was up with food. So we went over. 'Fritz' Maier ran over and said 'I just saw a damn Jap come up the hill over there and he ran back when he saw us.'" McCarten and Burtell went to find him. "I said to Larry that Son-of-a-B has got to be in that cluster of bamboo or that tall grass. Just as I was saying that I heard the snap of a Jap grenade. I hollered 'grenade!' and just as we were diving for cover I saw the thing go off, the black smoke and orange flame, and I got hit in the back." The Japanese had probably put the grenade to his head because there was nothing left of it after they got up and checked his body.

Burtell was able to walk to the aid station on his own. "I don't think I was hurt that bad. Doc Flannery, our old regimental surgeon, dug around on my back and said he didn't find much there, but he said 'Burt, you got a Purple Heart.' So that was that." After this, Burtell went on a few more patrols. He spent some of his time drawing sketches of things he had seen in the war. In one sketch he drew two drained, faded men, asleep in a shallow foxhole, their weapons in easy reach just in case. This and the others he would bring home, developing them into water colors and oil paintings that honor all of those who fought and died in this most brutal conflict.[11]

In the second half of May, several companies were pulled back for a rest. Oie had fled with what amounted to a bodyguard. Every bit of information gathered from prisoners and patrols showed that the Japanese left in the hills around Cuernos de Negros were hungry and sick. If Oie had elected to attack, he would have been slaughtered by American artillery and firepower. The U.S. 8th Army wanted to wrap up this fight and move on to Mindanao. The infantry was sent even further into the mountains to dig out the last remnants of the Japanese. The 1st Battalion's forces were pulled back for a rest and 2nd Battalion carried on.[12]

Thinking that Oie might try to escape to the west side of the island, Mahoney sent Company E by sea to Cauitan Point, to scout the Nagbalaye River. There they found very little except jungle, swamp, and a multitude of leeches. Returning to Dumaguete, the company went back into the hills for more patrolling. Sick and hungry Japanese soldiers could still kill a careless soldier. Zane Jacobs recalled a "cocksure 2nd looey" who joined E Company. He complained that the soldiers were dirty and unkempt. "We didn't have clean socks for about three weeks, what did he expect?" A few days later, as his platoon advanced, the lieutenant began to walk across an open clearing. Jacobs tried to warn him, but was brushed off. "He took about 10-12 steps and a .25 caliber shot got him in the gut. Chuck Walker came up and asked me for my sulfa powder and bandage to treat him. He got a one-way ticket."[13]

The Japanese could still fight. Company A ran into a position late in May and had to fall back from "heavy automatic and rifle fire." This may have been the fight when Gene Brinkman ran out under fire to rescue a badly wounded member of his platoon. Brinkman received a Bronze Star for it, but had mixed feelings about it in that the man had died in the end. G and F Companies fought a brisk series of actions to reduce another group dug in on steep slopes with well placed crossfire. In June, two platoons in Easy Company spent four miserable days reducing a "well constructed defensive position" on yet another ridge. They killed fifteen of the enemy but lost a man killed and seventeen wounded in the process. But there were better days, like this one recorded in F Company's log: "killed a Jap said to be a captain ... Patrol returned at 1200 [with no losses]. It has rained all day and everyone is wet and their foxholes full. The night was quiet with cold and rain."[14]

Claude Parish, a very young rookie on Negros, recalled three vivid memories of Negros more than six decades later. There was the night "all hell broke loose" because of a noise outside his unit's perimeter. "We let loose with a machine gun and found a dead cat there the next morning." Then the time the Japanese "shouted and worked the bolts on their rifles, to keep us awake." Finally, in the last days "we were waiting to go back, and singing a little song someone made up – 'there is no spaghetti here in Dumaguette, let's go back to Cebu.'" Parish got through the campaign without a scratch, but was annoyed when another man, who "cut his hand on a ration can lid," tried to get recommended for a Purple Heart. That bothered me. We were all scared and thought we'd get killed invading Japan." If someone had told him and his friends that the regiment was in its last battle, they would never have believed it.[15]

At long last, the 8th Army was satisfied that Oie no longer posed a threat. General Arnold ordered the 164th to turn Negros over to the 503rd Parachute Infantry Regiment and the Filipino guerillas. The two haggard battalions sailed back to Cebu on June 22. They had paid a heavy price to "secure" Negros. Thirty-three men had died and a hundred and seventy-nine had been wounded. Many men were suffering from dysentery, jungle sores, infections from leeches, and the ever-present malaria. Japanese casualties were given as 579 killed. A handful had been taken prisoner.[16]

While the Negros campaign ended, the 3rd Battalion's Company K was sent to Mindanao. There the men garrisoned a supply depot while other 8th Army troops struggled to eliminate the enemy on that island. Company K saw no combat but found the jungle on Mindanao so wet and thick with humidity that it reminded Paul Guzie of Bougainville. Already ill with yellow jaundice, Guzie "walked in my sleep one night, out to the edge of camp where a guard saw me and brought me back." The company medics examined him and thought he should ask for a transfer home, but he said no and returned to Cebu with his buddies.[17]

Back on Cebu, the men spent the last days of June in "improvement of bivouac areas and reconnaissance of Regimental Training areas." The Americal Division headquarters informed Mahoney that the division was being transferred to Lieutenant General Walter Krueger's 6th Army. They also ordered Mahoney to stage "amphibious exercises." These exercises involved organizing men into "boat echelons" and using wooden mockups of landing craft to practice a beach assault. Then, once real craft were available, "detailed training afloat" would begin. The men were told that the division was going to land under fire on the beaches of Japan. "Part of our training for the big invasion," commented Zane Jabobs, was being given information that "projected 75 to 90 percent casualty figures for our landing zone. Naturally we were all on edge."[18]

George Dingledy was now a sergeant for the 2nd Battalion. His friend, Charles Davis, was a sergeant for the 1st Battalion. Davis acknowledged that as administrative personnel they had had comforts that others envied. "We both had gone on some patrols on Bougainville. I remember I went on one to escort another patrol back in. I had the BAR, which you couldn't carry at port arms, let alone across you body. The trail was too narrow for that. On the way to the Torokina River, we saw a Jap patrol coming down toward us. We faded into the jungle and I stood on a little rise covering them until they went by, unaware of us. If they had seen us it would have been a mutual massacre." But neither man had been hurt so far. Now, at Cebu, both Dingledy and Davis were training men for landing on a hostile shore. "We were getting in some air force men who had no infantry training at all, and George and I had to start training them to be riflemen." Dingledy wrote a letter to his mother in early August in which he wrote about a number of casual things and described an elaborate staging of the musical "Oklahoma." Finally at the end of the letter, in regard to asking about a girl he had known in school, he wrote "could someone be thinking of post war? Not a bad idea at least."[19]

Another batch of experienced men learned that they had received furloughs and would soon be leaving for the States. The lucky ones included Doug Burtell, Bob Dodd, John Paulson, Alvin Tollefsrud, and Rudy Edwardson. Tollefsrud left with a mixture of relief and bitterness. "There were a lot of old timers that got killed around Dumaguette, they should have gone home, too." Edwardson left on a ship bound for Pearl Harbor, probably the same transport that carried John Paulson back. Paulson was happy to go, but felt bad about a little Filipino boy that he had been sharing rations with. "All his family was gone apparently, and he asked to go home with me. I couldn't take him and he cried. Other guys promised me they'd look after him. It's always bothered me that I never could find out what happened to him."

Burtell, along with his old friend Bob Dodd, got passage on a ship routed to San Francisco. Neither man quite believed that they were finally going home. Their voyage took the ship across the route of the *USS Indianapolis,* which was racing to a new assignment after delivering the first atomic bombs to the 20th Air Force on Tinian. The *Indianapolis* was torpedoed and sunk by a Japanese submarine just a few days later, with very heavy loss of life.[20]

The U.S. Navy's Task Force 38 struck targets in Japan in late July. Dive- and torpedo-bombers hit depots, warehouses, airfields, and ships in harbors. The Japanese put up little air opposition but anti-aircraft fire was intense and a number of fliers were shot down and killed. American battleships and cruisers bombarded industrial complexes at Hamamatsu a few days later. The Navy was almost unchallenged off the coast of Japan.

Admiral King, whose manipulations had kicked off the Guadalcanal campaign in 1942, commented that "at this time the Japanese Navy had virtually ceased to exist--we were fighting an island, not an enemy fleet." But there was no evidence that the enemy intended to give up. The Navy began to prepare for the invasion of Kyushu.[21]

Rolf Slen, the 494th Bomb Group navigator, was now stationed on Okinawa, which had been seized from the Japanese at a very high cost in American casualties. From Okinawa, Slen's bomber group attacked installations in Japan. This included one mission against the last floating Japanese battleship, the *Haruna*, which was already so badly damaged that it could not put to sea. Slen had been on thirty-five air missions so far. He thought he might get a furlough after completing forty or fifty missions, but also understood that he could remain in the Pacific until the war ended.[22]

Thousands of American soldiers, engineers, Seabees, and other personnel were also staged on Okinawa, all preparing for the invasion. A Minnesotan named Bernard Wegner, a member of the 6th Navy Construction Battalion, was there with his mates building camps for the hundreds of thousands of GIs expected from Europe. "We were told these men would be going to Japan and that the army leaders expected that when they invaded, they'd have to kill women and children in Japan, because they would fight us, too." Eugene Solomon, the lab tech stationed on Guadalcanal in 1944, was now stationed on Guam. "We were setting up a new hospital and they told us we would get thousands of badly wounded men once the fighting began in Japan. I think everyone expected a real bloodbath."[23]

But everything changed on August 6, when two B-29 bombers flew over the Japanese city of Hiroshima. Although Japanese radar operators had detected the planes, they took no action to intercept them; the Japanese had so few fighter planes left they decided to ignore "small plane raids." At 8:15 a.m. (local), the lead plane, the *Enola Gay*, dropped an atomic bomb on Hiroshima. Announcing this action in a press statement, President Truman warned that if the Japanese government "does not now accept our terms [of surrender], they may expect a rain of ruin from the air the likes of which has never been seen on this earth." But the enemy made no reply so on August 9 a second atomic bomb was dropped on the city of Nagasaki. Russia declared war on Japan a few hours later. Finally, Japanese emperor Hirohito bowed to the situation and broadcasted an announcement that Japan would surrender. Truman announced that Japan had capitulated. Pending the treaty signing and occupation arrangements, the war was over.[24]

News reached the Americal Division and ran like wildfire through the ranks. Arnold's headquarters made a stern statement that until something "official" happened, the landing training would continue. But this did nothing to prevent what the division historian called "bedlams of joy" as men ran in every direction yelling 'it's over' and 'thank you, Harry Truman.' The scene was repeated across the Pacific. Out in Leyte Gulf, ships fired off star shells and flares in an impromptu fireworks display. At numerous bases, men used anti-aircraft and machine guns to fire off tracers. Some celebrations went too far. At Rolf Slen's airbase, shrapnel from AA fire showered down and killed at least two men, prompting Slen and others to run for the bunkers. What an irony, he thought, that men were listed as 'kia' after dying in the victory party.[25]

At the 164th Colonel Mahoney cancelled orders for patrols to "seek and destroy" Japanese parties that were still hiding in the hills. Instead, patrols were sent out with interpreters to find the Japanese and see that they knew hostilities were ending. Each patrol was well armed, in case the Japanese refused to believe it. However, these missions came off without violence and on August 28, at a formal surrender ceremony in northern Cebu, 2667 Japanese gave up their weapons. After that, the Americal began packing all its equipment and preparing for transport to Japan, where it would become part of the occupation forces. Most of the 164th regiment arrived at Yokohama on September 10. Some would be there for only a few weeks, others for a year.[26]

No one knew what to expect from the Japanese. Would an enemy who had so recently fought to the last man now cooperate with their conquerors? Naturally suspicious, the GIs were very cautious their first weeks in Japan. Paul Guzie was part of a unit sent out to accompany Yokohama police as they confiscated weapons. He was leery to be among so many Japanese but quickly discovered that they feared him more. "All the houses were dark and shuttered because the citizens did not know what to expect [from us] and were afraid. We did see a few people peeking through the shutters." The GIs and police collected hunting rifles, large knives, ceremonial swords, "any other kind of weapon" and delivered them to a site that "destroyed all of them."

Guzie dealt with some Japanese who spoke English and could act as intermediaries. He gradually relaxed, finding the Japanese to be polite and cooperative and even, after a time, friendly.[27]

Medical personnel came into close contact with the Japanese before most other soldiers. Thousands of Japanese civilians had burns and other wounds from the bomb raids and illnesses that had not been treated because their nation's medical services had been overwhelmed. The American doctors and medics stepped in to help. Fred Drew was a young medic who had joined the regiment on Cebu in May of 1945. "I applied for pilot training in mid-'43 but the air war in Europe was winding down and there was not much need for pilots in the South Pacific." So he was trained as a rifleman and off he went for the Philippines. When he landed in Luzon, "one guy who seemed to be in charge told me I was a medic and to go with a red-headed sergeant, Art Bevore." When Drew protested that he hated the sight of blood and had tested as expert with the M1 rifle, Bevore gave him the standard Army answer: "Mac you are a medic and get your ass over here." So he was trained as a medic, and much to his surprise found he was pretty good at it. "I could hook up IVs which were tough and unknown to many medics." Drew helped with malaria cases and was "dumbfounded" to see B-17s being used at low level to spray DDT over the wetlands. As malaria continued to spread, he and others finally found that, as in the Solomons, mosquitoes were breeding inside coconut shells. He treated just one combat casualty, a soldier brought in with "shrapnel in the butt." He was finishing his medic training as the war ended.[28]

In Japan Drew accompanied a unit that was sent to a Japanese base to gather weapons. They met several unarmed "very sullen looking" officers who showed them to an aircraft hanger filled with "hundreds of rifles and swords and pistols." Seeing this Drew relaxed. He and the others had wondered "what was going to happen." The weapons hauled off, Drew was ordered to find a place to set up an aid station. Looking around he met an "old man" who bowed deeply, said 'hai' and showed him a kitchen/shower area he could use. The man was frightened at first but was so cooperative that Drew felt safe enough to holster his .45 pistol. Later a more formal medical station was established at one of Yokohama's hospitals.

Another time Drew was sent to a "comfort station" where prostitutes were kept by the Japanese Army for the convenience of their soldiers. "Several of them had VD which I treated. For a pretty young, naïve guy like I was, that was embarrassing." When the Americal Division demobilized, Drew transferred to the 98th evacuation hospital. He kept busy with sick soldiers and civilians, sterilized batches of needles, and became a regular 'doc' to the GIs. He also came to like most of the Japanese. He was fascinated by their formal manners and quiet dignity, and liked many aspects of their culture. He collected some Japanese art carvings and magazines, bringing them home when his tour was over, and keeping them for another sixty years. "I was a kid fresh from home and grew up a lot in Japan. It was a real experience for me."[29]

Others carried out similar assignments. Gene Scolavino performed a host of medical duties and, like Fred Drew, did inspections of prostitutes. His friend, Joe

Castagneto, became a company cook for the Americal. George Dingledy did a variety of things but found that the "duties were light and [we had] plenty of free time to explore the surrounding areas." He especially enjoyed a trip to Mount Fuji. The Japanese seemed to be cooperative in every way. But in the background, at the level of the emperor and his government, efforts were made to destroy incriminating records, especially records that could be used for war crimes prosecutions. In the months ahead American authorities would compile evidence to show that almost forty percent of the approximately 30,000 Americans who had been prisoners of war of the Japanese had died of malnutrition, mistreatment and outright murder. Some Japanese commanders would be convicted and executed for these atrocities, others would serve prison sentences. Memories of these things lingered long in the psyche of the Americans who fought in this war.[30]

Chuck Walker searched for weapons and assisted one group of men who were looking for documents connected to American corporate interests in Japan. When he and the last of the officers of the 164th were earmarked to return home, they moved first to the 43rd Division which would also be going back. While there, Walker was assigned to lay out a tent camp for an incoming division of new troops. Walker toyed with the idea of remaining in the Army but decided that its bureaucratic procedures in peacetime would drive anyone around the bend. Besides, he learned that if he did stay in, he would get no more than a short leave to the States and then return to Japan for extended duty. Forget it, he decided, he had a wife and little girl back home.[31]

None quite realized that, after the horrors of this war and despite the enemy atrocities and the eventual trials for war crimes, they were still the first echelon of American 'diplomats' in a process that was going to create a firm friendship between Americans and the Japanese. Most would not really have believed it if someone had told them that in 1945. To most every man the essential thing was that the war was over.[32]

The 164th Infantry was officially deactivated from Federal service on November 24, 1945. Any man in the regiment who remained in the Army to complete his service was transferred to another unit. With few exceptions the men were already counting the days before they accumulated enough points to go home. But a lot of them worried about going home, too. How much had it changed? How much had they? How would they live in a nation at peace? The only way to know was to go home and find out.[33]

(Endnotes)

[1] Milton Shedd to his wife, letter contained in a bound collection of Shedd letters, kindly shared with Shoptaugh by his son William Shedd; George Dingledy to sister, letter copy shared with Shoptaugh by George Dingledy.

[2] Dingledy letter to sister, April 22, 1945.

[3] All the plans and casualty estimates are discussed in Richard B. Frank: *Downfall: The End of the Japanese Imperial Empire* (New York: Penguin, 2001).

[4] American projections of Japanese resistance were if anything too optimistic. Frank provides Japanese sources showing that Emperor Hirohito was prepared to sacrifice hundred of thousands of his subjects to obtain a negotiated peace, while hardliners in the Japanese military thought "national suicide" preferable to surrender. D. M. Giangreco, *Hell to Pay: Operation DOWNFALL and the Invasion of Japan, 1945-1947* (forthcoming from Naval Institute Press) promises an re-examination of the whole invasion debate using newly released documents.

[5] Two indispensible works in connection with U.S. negotiations with Russia and the development of the atomic bomb are Leon V. Sigal, *Fighting to a Finish: The Politics of War Termination in the United States and Japan, 1945* (Ithaca, NY: Cornell University Press, 1988) and Richard Rhodes, *The Making of the Atomic Bomb* (New York: Simon and Schuster, 1995).

[6] Robert Love, telephone interview, January 23, 2008, supplemented by Love e-mail to Shoptaugh, January 27, 2008. Love joined the regiment in Leyte and survived the battles on Cebu and Negros without wounds, but contracted malaria which he "toughed out" first in his home town of Thief River Falls, Minnesota, and later in Florida.

[7] Clauson wrote his account of the "Last Day in Combat," in the 1990s and told the story again in the video interview with his nephew, January 28, 2007. The entire written version is in the *164th Infantry News*, July 2008. See also the S2 Journal entries for May 6-8, 1945 for accounts of B Company's attacks and losses.

[8] G and F Company actions in S2 Journal entries for May 6-7, 1945; "Company F, 164 Infantry Log," May 6, 1945.

[9] "Company F, 164 Infantry Log," entries for 8-11 May, 1945. Cooper and Smith, *Citizens as Soldiers*, pp. 322-23, summarize the last phases of the Negros fighting. See also Cronin, *Under the Southern Cross*, pp. 323-31.

[10] S2 Journal entries May 18-19, 1945.

[11] Burtell interview with Shoptaugh, October 12, 2007 and Burtell's written memoir. Some of Burtell's sketches are reproduced in the March 2008 issue of the *164th Infantry News*.

[12] S2 Journal entries suggest that a reason given for pushing further into the Cuernos de Negros hills was to protect the Negros airstrip from Japanese mortar fire. The strip held lighter spotter planes. All the combat fighters and bombers were stationed on Leyte and Cebu.

[13] Walker, *Combat Officer*, pp. 218-21, describes the frustrating reconnaissance at Cauitan Point. Zane Jacobs story in e-mail to Shoptaugh, March 5, 2007, with additional detail in Walker, pp. 215-16.

[14] S2 Journal entries for May 20-22, June 8-11, 1945; "Company F, 164 Infantry Log," entry for 18 June 1945. Even sixty-five years later, the aggravation that men had for this slogging combat on Negros is easily sensed in the S3 "Operations Report, Cebu/Negros."

[15] Claude Parrish, telephone interviews with Shoptaugh, November 24 and December 9, 2007.

[16] S3 "Operations Report, Cebu/Negros.," entries for June 20-22, 1945.

[17] Cooper and Smith, p. 323; Guzie interview with Shoptaugh, July 7, 2007.

[18] S3 Operations memorandum, July 25, 1945, 164th Infantry Association Records; Jacobs e-mail to Shoptaugh, March 5, 2007.

[19] Charles Davis, telephone interview with Shoptaugh, October 7, 2007; Dingledy letter to his mother, August 2, 1945, copy courtesy of George Dingledy.

[20] Tollefsrud interview with James Fenelon; Edwardson interview with Shoptaugh; Paulson interview with Shoptaugh; Burtell interview with Shoptaugh; Robert Dodd Jr., *Once a Soldier,* p. 78.

[21] Admiral Ernest J. King, Commander in Chief, United States Fleet, and Chief of Naval Operations, "Third Report to the Secretary of the Navy, Covering the period 1 March 1945 to 1 October 1945," (http://www.ibiblio.net/hyperwar/USN/USNatWar/USN-King-3.html). See also Samuel Eliot Morison, *Victory in the Pacific, 1945* (Boston: Little Brown, 1960), pp. 330-31.

[22] Rolf Slen, interview with Shoptaugh, July 29, 2007.

[23] Bernard Wegner (not to be confused with Bernard Wagner of the 164th), telephone interview with Shoptaugh, June 21, 2007; Shoptaugh telephone interview with Eugene Solomon, April 24, 2007.

[24] For an incisive examination of how the Japanese decided to surrender, see *Japan's Longest Day* (English edition, Kodansha International, 2002).

[25] Slen interview with Shoptaugh.

[26] S3 orders and reports for August 14 and 16, September 10, 1945; Cronin, pp. 348-52.

[27] Guzie, "From New Caledonia to Japan, 1943-1946."

[28] Dr. Fred Drew, "Four Comrades in My Young Life," *Americal Division Newsletter,* October 2006, together with telephone interview of Dr. Drew by Shoptaugh, June 28, 2007.

[29] Fred Drew narrative (March 2008) given to Shoptaugh, and Drew interview with Shoptaugh. After leaving the Army, Drew studied medicine and became a podiatrist. The Japanese artifacts he collected during the occupation, together with the narrative including several photographs, are now being added to the 164th Infantry Records at the University of North Dakota.

[30] Castagneto/Scolavino interview; Dingledy, "I Remember." For Japanese war crimes, consult *Hidden Horrors: Japanese War Crimes In World War II* (Boulder, CO: Westview Press, 1997), by Japanese historian Yuki Tanaka. The numbers of American military prisoners is imprecise because Japanese records were incomplete, but the estimate of death rates for American prisoners is 37 percent. Some 14,000 American civilians were interned by Japan and suffered cruelly as well.

[31] Walker, pp. 233-36.

[32] John W. Dower, *Embracing Defeat: Japan in the Wake of World War II* (New York: W.W. Norton, 1999), is the essential study for the occupation of Japan and the forging of American-Japanese friendship.

[33] Cronin, p. 386, describes the deactivation of the Americal Division and its components.

16: Comrades

Over sixteen million American men and women served in the nation's armed forces during the Second World War. That was more than twelve percent of the American population at that time. But this common tie did not guarantee that any one veteran could automatically understand any other veteran. Such was the nature of a mass, mechanized army that only about one man in seven served in combat. Those who did strongly believed that non-combatants could "not relate" to their experiences. But it went even deeper than that. Because the war had spanned the globe and was fought in every kind of terrain and weather, on the land, sea and air, the combat veteran often also felt strongly that only someone who served in his war -- *his* theater, *his* battles -- could really *know* what he went through – the common experience was what made soldiers into comrades.[1]

The 164th had begun as a National Guard regiment. Its motto "I Am Ready" (Je Suis Pret in the original French) reflected the tradition of the old militia, the 'citizen soldier' who was ready with his musket to defend his nation. But the war itself made clear that in contrast to the individualistic imagery of the citizen soldier, it was mass organization, mass fire, and unit discipline that won the victories and overwhelmed the fighting power and will of another modern state. In this sense *"They are Ready"* better fit the Second World War, when large numbers of well-trained troops taken from all parts of the nation, all ranks of society, learned the ways to win not as individuals but as units. Group integrity played a large part in how the men fought their war together, won it together and related to one another in the years after it ended.²

At Guadalcanal, the glue that held the men most tightly together was their common experiences, common localities and common backgrounds, when neighbors and friends fought alongside one another. After 1942, the annealing factor, as the regiment was continually drained by casualties and reformed by replacements, came in the comradeship forged in combat -- the veteran's commonsense recognition that his life rested in the hands of the man alongside him. Nothing better reflected this than the constant concern among the men on the abilities of their squad and platoon leaders. No one better expressed the ties that the war forged than Gerald Sanderson: "my closest friend came out of this [war], you can't get better friends than those you risk your life with."³

The 164th's wartime record after Pearl Harbor was significant: one of the first units sent to the Pacific after the war began, its men had participated in five campaigns and dozens of battles stretching over more than 600 days in contact with the enemy. Men in the unit were awarded six DSCs, 89 Silver Stars, 199 Bronze Stars and over 1200 Purple Hearts, making it one of the most decorated outfits in the Pacific theater. ⁴

A number of the longest serving members of the regiment returned to their North Dakota and settled again in their home towns. Usually they made this choice because of family or employment, but several chose to remain in the northern climate because doctors had warned them that recurrences of malaria would be more frequent in the southern or western states. Many men took that advice. Virtually all of them had relapses over the years. Rita Fox, the daughter of Murphy Fox, vividly remembers her dad: "every few years, he would be shivering under blankets on a warm day, and Mom saying 'it's okay, he will be okay'." Fox bought a farm in the early 1950s and liked to discuss the farm markets with his old sergeant, John Tuff.

When they first came home most men wanted to get on with their lives and put the war behind them. The Army gave veterans a small booklet entitled "Going Back to Civilian Life." It contained information about the GI Bill of Rights, getting medical assistance from the Veteran's Bureau, some other things. And it gave the returning GI some advice: put the war behind you; get on with your life; forget it as best you can. If a man had "personal problems," then "an officer will be found at most posts . . . who is there to help you and your dependents after your separation." While this sounded good and while the Veterans' Administration did as much as their budget and personnel sustained, providing service to millions of men was a difficult and expensive task.[5]

Once peacetime production began again, finding a job was relatively easy. United States was the nation that had won the war with relatively few casualties (300,000 killed, compared to 10-15 million in Russia, four million in Germany or nearly three million in Japan). Whatever cost the dead had on family and friends, the nation did not suffer the economic destruction that Russia, Germany, and Japan did. Instead the nation's economy markedly improved. The great depression of the 1930s was washed away in a tidal wave of war production, around-the-clock factory work, and rising wages. Unemployment was almost nonexistent. The gross national product rose from $88 billion in 1939 to $135 billion in 1944. People suddenly had money in their pockets, sometimes for the first time ever. Personal savings, encouraged by wartime rationing and publicity campaigns, grew at an unprecedented rate. Eighty-five million Americans bought billions of dollars in war bonds. The good times were rolling again, and in the marriage and baby boom that followed the war, with the consequent housing, suburban, and education booms of the 1950s, America entered into an era of unprecedented prosperity. The American economy dominated the world markets for the next quarter century and the average standard of living essentially doubled in a decade.[6]

Once home, the ex-GIs claimed their share of this bounty. The majority of those who left the military were more than happy to be civilians again, pursuing jobs and careers and families. Quite a few of the men used the GI Bill of Rights to resume their education. George Dingledy and Fred Drew were among those who went to college by using the Bill's provisions for education assistance. Dingledy got home with "$118.70, the most money I had ever had at one time in my life," and could get the twenty dollar per week payment for one year (the "52-20 club"), which he used "mainly for beer." But a college education had always been his dream. The GI Bill paid a full tuition for up to four years. Dingledy entered St. Joseph's College in Rensselaer, Indiana, in the spring of 1946. He worked odd jobs through college and after graduating went to work in appliances, ultimately owning his own business. With the education aid, Jim Fenelon went to college and then to law school. Fred Drew got home and used his GI Bill benefits to study medicine. He became a podiatrist, delighted to be using his medical skills in peacetime practice. Between 1945 and 1956 some eight million World War II veterans went to public and private colleges with Federal help.[7]

Most others found jobs. Phil Engstrom, after completing his rehabilitation for his wounds and being discharged in late 1943, got a job with the U.S. Postal Service. Walt Rivinius, who had been wounded with Engstrom at Koli Point, was likewise discharged, used the GI Bill for training and went to work repairing organs for a music company. Horace Nearhood worked back home at the Jeep plant for few months and then became a bricklayer. John Tuff returned home and set out to pursue his dream to farm. "He loved farming," his wife noted, "but also worked at a service station and at Rugby Motor Company for ten years, to add to our income." Rudy Edwardson worked at a couple of jobs and then bought a motel in Carrington which he ran for nearly three decades. Alvin Tollefsrud got a job as a bookkeeper for a bank in Mayville and went on to become the bank's president before he retired. Milt Shedd became a stock broker. A lover of all things oceanic, he played a major part in creating a nonprofit marine research foundation. He also helped build the Sea World park in Florida and served as its board chairman. Each man in his own way grabbed a piece of the dream that he had fought to protect.[8]

Some found the transition from war to peace very difficult. Paul Guzie later noted that he "was a young 18 when I entered the service and came home a very mature 21. When I came home I was surprised to see the changes in attitudes and life styles of my friends and neighbors. I realized then that things would not be the same again." He was troubled by what he had seen and some of what he had done in the Pacific. He found he could not sleep in his old room at his parents' house. It was too enclosing, stifling. He began sleeping on the back porch, preferring the fresh air and openness. But any little noise or change could bring him to a state of instant alertness. "I could tell when a noise or smell did not belong where I was." This almost led to tragedy. "In the middle of the night I heard a noise in the basement. Half awake, disoriented and acting on battle-honed instincts, he "got a knife and started down the stairs." The noise was his father, returning from a late work shift and changing his clothes. Paul recognized him at the last second. "I was shook up because I had come so close to killing him." He asked his parents to be careful not to surprise him. They in turn urged him to find someone to talk to about his war memories.

Guzie found help at a settlement house in Minneapolis, where he met a young woman who was working there as a social worker. She listened to him as he related his experiences, gave him some advice and encouragement – and later married him. Several other men had problems such as Guzie's, but this was not a time when counseling was a readily accepted part of the culture, and very few sought help as he did. Many groped for a way to treat themselves. Not a few tried to do so through alcohol. "Dad never said much about the war, but it bothered him and he drank an awful lot for a long time," said the son of a man who joined the regiment on Bougainville. "Finally he got some help and quit, but the war was always there in the back of his mind." There were other men who quietly admitted in confidence that they drank more after the war than before it."⁹

Not every man chose to put his military experiences behind him. Some decided to stay in the military or rejoin it. Mostly they were men who saw the career potential. John Stannard, whose stand on Guadalcanal helped get him into West Point, graduated in 1946 and spent the next three decades in a number of assignments, retiring as a brigadier general. Dick Stevens stayed on and rose to the rank of colonel. Fred Flo became a brigadier general, and Charlie Ross commanded combat soldiers again in Korea and Vietnam before retiring as a Lieutenant Colonel. Among enlisted men who opted for a military career was Lester Kerbaugh who tried farming after the war, but "didn't like it" according to his son. He reenlisted in 1947 and went to war again, in Korea, where he earned a battlefield commission, a second Purple Heart and a third Bronze Star. He did not retire from active duty until the late 1960s. Dick Stevens went home and spent most of his first thirty days in the states resting in a good hotel in Hot Springs Arkansas. "We went to USO shows and had good times." Then he met a "sweet girl" on a blind date that he took a liking to: "I wanted to see her again so I pulled a string or two and got an assignment as an office NCO. About the time she agreed to marry me I decided to reenlist." Stevens later went back overseas, to postwar Germany. "I became a platoon sergeant in a military police company in Berlin. Then I got into the OTC program and got a commission." Stevens served in Korea as commander of the 25th Infantry Division's MP company. During the Vietnam conflict he acted as a liaison to the commander of all the South Vietnamese military police. By the time he retired, Stevens was a colonel. "The Army was good to me." Both Wendell Wichmann and Al Wiest, Stevens' commander on Guadalcanal, also stayed in the Army. Both also retired as colonels.

Joe Castagneto left the service for a few years and then in 1948 he, too, decided to enlist again. His first assignment was with an honor guard unit in Brooklyn that escorted the remains of recovered war dead from the docks to the railroad stations. Joe stayed in the Army for some thirty years, retiring as a sergeant major. Castagneto's friend, Gene Scolavino, also returned to the service, joining the newly independent U.S. Air Force. In 2007 they met again at the regimental reunion.

Other men came home but joined the Army Reserves or kept their standing in the National Guard. George Dingledy joined the reserves and in 1950 was alerted for possible active duty. "I had only a few months left and with a wife and kid they decided not to send me to Korea, but I stayed in the reserve another year." Bernie Wagner remained in the Guard and was called up for active duty during the Korean War. He went to Korea and was put in command of a training unit. When a group of guardsmen left Rugby, North Dakota for Korea, a comrade of John Tuff told Tuff's wife Edith that he was happy John was not among them. "I can still see him as he ran zig-zagging across the ground [in the Pacific], a couple of bullets hit his canteen and cut through his belt. I'm glad he's not going to Korea, he's done enough." Woody Keeble, however, did go to Korea and there he fought with the same intensity that he had shown in the Pacific. In 1951 he destroyed a series of Chinese bunkers while wounded, saving the lives of several men in his unit. He eventually received the Medal of Honor for this.[10]

Father Thomas Tracy had been awarded a Bronze Star (for valor) during his Pacific service. Still with the Guard in 1950 he went to Korea and again went up to the combat zones to be with the men. Tracy never failed the men, commented his old friend Harry Tenborg. "He repeatedly disregarded his own safety to offer guidance and encouragement to the troops." Tracy retired from the Guard in 1954, holding the rank of a major. His heart, weakened by malaria, plagued him in his later years. When he died of heart failure in 1960, at the age of forty-nine, his 164th comrades mourned the loss of their "padre of the foxholes."[11]

With prosperity and peace, Americans wanted to put aside the sadness of the war. During the war books about it, its causes and consequences, and the nation's role in the postwar world had proliferated. But after the war it was different. Aside from a few memoirs by prominent military leaders, such books did not sell well.[12]

Postwar publications advised the men and their families to treat the war as a closed book. Advice from a columnist for a news bureau that sold filler stories to hundreds of small-town newspapers was pointed: "Make up your minds, you mothers and sisters and wives everywhere -- the boys are coming home cross, vague, restless, critical, dissatisfied. He'll start up, to answer your questions, from some dark dream. 'What? What'd you say, Mom? Yep, we had pretty good chow at Guadalcanal. Nope, it was kind of rotten -- oh, I guess it was pretty good.' He's got long memories to live down. Give him plenty of time."[13]

Memories to "live down" – the words suggested something best forgotten, perhaps even shameful. Many relatives did not in truth want to know what their sons and husbands and brothers had experienced. A large part of the post-war culture was, in essence, urging men to regard the war as an interruption in their normal lives. But the war left an indelible imprint on most men and on their families:

> On December 7, 1945, four years to the day since Pearl Harbor, Bernard Scheer sat down in his home in Arlington, Minnesota, took some paper and wrote out in long hand about ten pages summarizing his war. He related how he and his friends enlisted in the Army at Fort Snelling, how he joined the 164th at Camp Claiborne and went with the unit to the Pacific. He wrote of the horrors of Guadalcanal, the way he and others buried the Japanese corpses after the Battle for Henderson Field. He related the numberless patrols on Bougainville and the fighting on Leyte before he was allowed to go back to the States on rotation points and returning to his parents' farm. "I remember greeting them and then getting a lump in my throat and actually I shed a few tears, which was unusual for me. I feel that my own prayers, the prayers of my parents and brother and sisters and friends, plus a lot of genuine luck, are the only thing that could have saved me. So many of our buddies were not so fortunate."

Finishing his statement, he folded the papers and put them away where they remained until his son, Dan, found them. "I had no knowledge of this narrative as it sat in his filing cabinet for 57 years," the younger Scheer wrote in 2007. "Dad is like most WWII veterans - he said very little about his war experiences."[14]

After more than three years in the Pacific, Doug Burtell sailed into San Francisco Bay on July 4, 1945. He was barely twenty-one years old. "At Fort McDowell they took away a lot of our souvenirs, but shipped them to me later. They gave us clean uniforms and we went over to the PX and got drunker than hell." Back in North Dakota, he finished high school and tried college, "but I was sitting there with 18 year olds and I just couldn't adjust." He worked at several jobs before settling into construction, took some classes in architecture and art, then went to work in the home improvement business and became the sales manager and part owner. He met a girl back home and married her in 1946. "I thought about re-enlisting a time or two, but I had a wife and kid." Even though the war was becoming more distant every year, he never considered discarding the drawings he made of the Pacific. It was after he retired that he took them out again to paint watercolors and oil paintings.[15]

James C. Simpson never made it home. Drafted in 1943, trained in Texas and sent to Leyte with many other replacements, he joined Co F and was killed on February 21, 1945, after barely a couple of hours in combat. His son, Dan, was born three months after he went overseas and never knew him. Dan's mother, Ruby, told him a little about his dad while he grew up, but one of Dan's most vivid memories was when Ruby remarried in 1958, "so she burned them. I remember her crying and putting the letters in the burn barrel." Years later, Dan asked the 164th veterans for any information about how his dad died.[16]

After the war, the Army's Quartermaster Corps recovered the bodies of over 280,000 war dead from the temporary burial sites in Europe and the Pacific and began transferring their remains for re-interment in permanent military cemeteries. When families opted to bring their loved ones home for burial, the man's comrades frequently asked if they could provide an honor guard. Rilie Morgan's father had his son's body brought back soon after the war and had him buried in the family plot at Grafton. The service was attended by several who had served with him. When Miles Shelley's family brought his body home in 1947, his cousin, John Paulson, and most of the other 164th veterans in Carrington formed an honor guard to accompany the casket from the train station to the cemetery. As Paulson later said, "Miles looked after us over there and we owed this to him."[17]

Sue Tolliver-Pompa's dad, Ernest Tolliver, died in the year 2000. "I'm sorry to say that's when I began searching for his service in World War II. In my search to learn what he did and why he didn't talk about his military service, I've come across small amounts of knowledge." Tolliver's brother persuaded him to do a very brief interview about ten years before he died, but Tolliver was very hard of hearing by then and had been weakened by illnesses, so most of what he said was general information about being in Company A on Guadalcanal and how poor the conditions were on the island. At one point he referred to call off an attack because "the Nips were coming in" and that the ground they gave up had to be regained later. "It was the costliest thing that our regiment ever had, our company lost many men in one day's time trying to take that same ground back, but we did it." The obvious references to Koil Point and the Matanikau were about the only specific information he recorded, and his daughter noted he "paused and looked so sad" when he said it.[18] Sad memories were common:

> "Swede" Clauson went home from the war, and returned to college at Hamline in St. Paul, Minnesota, where he majored in philosophy and religion. Working odd jobs while he took classes, he completed his degree in 1949; "he always joked that it took him 13 years to graduate from college," his sister later commented, "but he finished under the GI Bill." Clauson later lived in California, working for several years for a shoe company and then at Sears Roebuck. Eventually he returned to Minnesota and worked at Dayton's Department Store. The impact of the war was evident not only in the scars he carried from his face wound at Guadalcanal but also in his determination to champion the cause of peace. He was active "in the Nuclear Freeze movement and the [protests against] Honeywell," his sister remembered. Swede was willing to tell his war stories later in life but always made a point of saying to those who would listen that "war is a terrible thing."[19]

Almost no one could just forget; once a man had lived through the war, he had to live with the war. His family had to live with the war, too, and as the actions of Jim Simpson's widow suggest, that could be very difficult; her burning of his letters after remarrying gives some idea of the pain she felt and the price Simpson's family paid in what was later called the "good war." However great the victory was, the regiment had lost 325 men killed and 1193 wounded in the war. When Fred Hales, a Company L rifleman, gave an interview to Jim Fenelon forty years after the war, he said that each time he recalled the battles he asked "why didn't I get shot?"[20]

As Doug Burtell admitted, "you had dreams of the war and there was hardly a day when you wouldn't think something about it." And as Charles Walker pointed out, "I think I killed as many as a hundred men in the war. That's a part of you for life."[21]

Such memories were not easily shared. Bernard Scheer was not prepared to tell his parents, or later his wife and children, about burying Japanese bodies in 1942. Most men would not talk about the brutal fighting, the take-no-prisoners mood that prevailed on both sides, the death of mates that had become like family to them. "At some point I stopped asking about my dad's military service," Sue Pompa admitted. "The comment I remember hearing was 'a girl doesn't need to hear about those things.'" Even after Sue married and had her own children, her dad kept his silence. When her son asked Tolliver about the war for a school project, "he said he was at Claiborne on December 7, 1941, and he hated the Japs for that and it took him many years to get over it. That was it, not a word more."[22]

Combat's emotional wounds were salved by the veterans trust in one another. The 164th Infantry Association was formally established in 1946. The first president was Harold Zumpf, a member of Company H out of Jamestown. The first action of the Association was to plan and hold a reunion, which went off so well that the members decided to make it an annual event. Even the most tight-lipped man could generally talk about the war with one of his comrades. "I think only those who were there could really understand what it was like," Bernie Wagner commented years later. The highlight of each reunion came when company groups met separately to share memories. Part merrymaking, part history and part informal therapy, the reunions grew more important each year. The Individual company reunions were also held for decades.[23]

The Association's newsletter developed gradually, starting as an occasional one-page mailing of information about upcoming reunions, announcements of births and deaths, other miscellaneous items. By the 1950s the newsletter was growing into a more elaborate publication. Members were beginning to submit reminiscences in addition to announcements about family and career. In 1963 the Association invited Joe E. Brown to attend its annual reunion. When Brown accepted, the reunion drew a record number of veterans from outside North Dakota.[24]

By the late 1960s, the newsletter was becoming a vehicle for establishing the unit's history in the war. The memories that the men shared in the *News* reflected a growing trend in military history in which war was portrayed not from the "general's viewpoint" but the common soldier's experiences. The commercial success of such books as Cornelius Ryan's *The Longest Day*, John Toland's *But Not in Shame*, and Walter Lord's *Day of Infamy*, based on extensive interviews with dozens of ordinary soldiers and seamen were inspiring a "ground up" method for understanding what happened in the Second World War. The tape recorder, and later the video camera, permitted the men who did the fighting to relate their experiences as never before.[25]

Thousands of books on the war had been published by 1980. But even so, historians and popular writers still had written very little about the Army's role at Guadalcanal. Historian Fletcher Pratt's 1948 study *The Marines' War* acknowledged the 164th's stubborn stand at the Battle for Henderson Field but gave little detail. Furthermore, Pratt's reference to the soldiers as "big, burly men, a little slow on the uptake," made him few friends among the GIs who had been there. The official history of the Marine Corps, the Army's official histories and the Infantry Journal history of the Americal Division, while very good expositions of the Pacific War from the command perspective, provided only very limited looks at combat from the ordinary soldier's viewpoint. The newsletter therefore offered several veterans a chance to get their stories on the record. The publication of *Citizens as Soldiers*, a study of the state guard's history, did make use of interviews and provided a much needed correction to the story.[26]

The passage of time encouraged men to speak out more than they had in the first few decades after the war. "It is sad to realize that in a few years the WWII era 164th members will be gone," a member wrote in the 1980s.

"In the not too distant future, there will be no survivors." That feeling prompted stories, either as letters to the editor or in the form of articles commemorating anniversaries from the war. The stories were by turns funny, sad, whimsical, angry. Commenting on a photograph of a grave stone at the National Military Cemetery in Hawaii, a member wrote of the man buried there: "Sgt. Clyde G. Morgan was member of Company F … I can remember like it was yesterday that on November 10, 1942, Sgt Morgan and his men were ambushed by considerable numbers of the enemy and subjected to machine gun fire. Morgan was killed along with one of his men, named Hall. His second in command and two others were seriously wounded. The remaining members of the patrol withdrew and crossed jungle and a deep stream with the three wounded, undoubtedly saving their lives. I think when the history of the 164th is written Morgan and his men should get a couple of pages." "I noticed in the last issue of the *News* that George Winter had died [in 1989] and remembered that over in Bohol, George got enough points to leave," wrote another. "The day he was supposed to leave his buddy and mine, Sam Wheeler, got killed on a ridge. George took that pretty hard. Thinking about that a thousand times, these were crappy little deals that several good guys were lost in."[27]

Another role that the 164th Association took on with increasing regularity was helping with requests for information from family of deceased veterans. Dan Simpson learned more about how his dad died on Leyte because Crescencio Cruz read Dan's appeal in the *News*. Sue Pompa got a few bits of information and a few news clippings in the same manner. "My father died about three years ago without ever telling me about his time in WWII" one son wrote plaintively in an issue of the *News*. "Does anyone know anything that could help me and my brother and sisters in knowing more about my father?" For each such appeal there was usually a reply.[28]

After the mid-80s members of the Association raised money to help build a regimental memorial erected near the state capitol. They mounted another fund drive for a memorial to be built on Guadalcanal. Both memorials were dedicated in the 1990s. The Guadalcanal monument was dedicated in a 50th anniversary ceremony held on the island, attended by veterans who made the long trip, walked again over the ground and retraced the sites of the battles of their youth.[29]

The strong bonds within the Association grew stronger as mortality became ever more evident. Diagnosed with cancer, one veteran wrote that he did not fear death, having grown up "in a world that I felt could have been more loving, a little more caring." He had lived a good life after 1945, had a good family, and above all had the memories of his comrades – "while I never had a brother, I have always felt that Company F satisfied that . . . thank you for the friendship we shared." Similarly, when John Tuff became very ill in the 1990s, he made a point of calling on old buddies in the service. "John Tuff came to my dad's 80th birthday party in 1998," Rita Fox recalls. "He was very sick, but he insisted on coming with his wife Edith up to Dad's lake cabin for the celebration. John had helped him get his promotions in the service. I watched him and Dad that day, the way they talked to and even looked at each other. It was special bond, you could see that. Dad always said John was the bravest soldier he ever knew. And then my cousin's daughter asked them how they knew one another it. They said they'd been in the war together and she asked which one! Then when she asked if they had seen some action, John looked over at my dad and then said 'Oh, we saw a little.' John died not so long after that. Dad talked often about John and the others in his last years, up to his own death in 2004."[30]

Thurston Nelson paid a heartfelt tribute when Samuel Baglien passed away in 1989: He "has gone to a new duty station where it never rains on parade days and the band is led by Souza." After Baglien lost a lung to illness and was told not to fly, "he rode a bus to our reunion. It's going to be lonely not hearing from him at Christmas. He was an officer and a gentleman that those of us who are left will miss."[31]

The number of surviving members the 164th Infantry was dwindling. Several companies had only a few men able to attend the annual reunions. In 2007, Bill Tucker called his old captain, Al Wiest, and asked if he would be going to the reunion. "The Skipper said, 'you're damn right I am, nobody's going to stop me.' I came to see him. You know I wouldn't want to go through it again, but I wouldn't take a thing for the experience I had with those guys." By then there were fewer than six hundred members of the 164th Infantry Association and more than half were family of deceased veterans. But some men vowed that they will keep coming even if they have to be carried to it.[32]

Sadly Bernard Scheer would never be able to see his comrades again. Afflicted with various health problems, Scheer by 2007 had lost much of his memory due to dementia. His son admitted that he could not say for certain how much his dad understood "when I told him I was giving a copy of his short war memoir to the Association, but "when I mentioned the 164th, he smiled and said he felt 'very proud.'"³³

(Endnotes)

[1] Comradeship and what it meant to the soldier form the heart of three major studies – John McManus, *The Deadly Brotherhood: The American Combat Soldier in World War II* (Novato, CA: Presidio Press, 1998); John Ellis, *The Sharp End: The Fighting Man in World II* (New York: Scribner's, 1980); and Gerald F. Linderman, *The World Within War: America's Combat Experience in World War II* (New York: Free Press, 1997).

[2] The importance of mass armies in the industrial age is well covered in John Ellis, *Brute Force: Allied Strategy and Tactics in the Second World War* (New York: Viking Books, 1990).

[3] Gerald Sanderson interview with Shoptaugh. In a conversation at a regimental reunion over fifty years after the war, the only remaining member of John Paulson's squad said to him, "John, many times we wanted to tell you something and now I'm the only one left to say it – we thought you were the best squad leader we ever had, you protected our lives and ran the same risks as us. Thanks." That, Paulson said in 2009, was "worth more to me than any medal."

[4] Cooper and Smith, *Citizens as Soldier*, p. 328. The regiment also had the unique honor of Colonel Robert Hall's Navy Cross, the only such navy award to a soldier. It is also of note that nearly half of the total awards for the regiment were given at Guadalcanal, reflecting the parsimony that the Army showed later in the war when many acts of heroism went unrecognized.

[5] *Going Back to Civilian Life* (Washington DC: Army and Navy Departments, August 1945 edition), pp. 8-9.

[6] For two fine studies of the economic impact of the war, see Geoffrey Perret, *Days of Sadness, Years of Triumph* (New York: Coward, McCann & Geoghegan, 1973) and Richard Lingeman, *Don't You Know There's A War On?: The American Home Front, 1941-1945* (New York: Putnam, 1970). The best study of returning servicemen is Mark Van Ells, *To Hear Only Thunder Again: America's World War II Veterans Come Home* (Lanham, MD: Lexington Books, 2001).

[7] Edward Humes, *Over Here: How the G.I. Bill Transformed the American Dream* (Orlando: Harcourt, 2006). Among Humes's many findings is evidence showing that the original G.I. Bill returned seven dollars to the economy for every one dollar invested.

[8] "Walt Disney of the Sea," *164th Infantry News,* October 2009. Shedd's leadership in Sea World is noted in numerous articles and books.

[9] Guzie, "From New Caledonia to Japan, 1943-1946," together with Shoptaugh interview of Guzie, July 7, 2007. The son whose father drank asked for anonymity in a 2007 telephone interview with Shoptaugh.

[10] Wagner telephone interview with Shoptaugh; Edith Tuff interview with Shoptaugh. Keeble's Korea exploits are described in Merry Helm's new book, *Woodrow Wilson Keeble: The Man Called Chief* (Fargo, ND: privately published, 2009).

[11] "Father Tracy, Foxhole Padre, Dies," Fargo *Forum*, October 21, 1960.

[12] On the postwar economic boom, see Donald W. White, *The American Century: The Rise and Decline of the United States as a World Power* (New Haven, CT: Yale University Press, 1999), pp. 57, 143, and David M. Kennedy, *Freedom from Fear: The American People in Depression and War, 1929-1945* (New York: Oxford University Press, 2001), pp. 852-58. Winston Groom, reviewing Alastair Cooke's book, *The American Home Front: 1941-1942* in 2007, noted that Cooke had finished the book in 1942, but "no publisher was interested" because few would want to read about the war years after it was over.

[13] Kathleen Norris, "The Boys Come Back Changed," *Moorhead Daily News*, July 29, 1944. Norris wrote advice columns, short fiction, and editorials for the Western News Syndicate.

[14] Bernard Scheer, "Resume of My Army Career"; Dan Scheer, e-mail to Shoptaugh, March 8, 2007.

[15] Douglas Burtell, interview with Shoptaugh.

[16] Dan Simson, telephone interview with Shoptaugh, September 28, 2007. For James Simpson's story, see the *164th News,* July 2006.

[17] The Army cemetery program is explained in the U.S. Army Quartermaster Foundation web site: http://www.qmfound.com/honors.htm. The American Battle Monuments Commission (www.abmc.com) also has a searchable database for locating a specific gravesite. Morgan information from John Hagen tapes; Shelley honor guard from John Paulson, conversation with Shoptaugh, July 17, 2009.

[18] Susan Tolliver-Pompa to Shoptaugh July 17, and September 22, 2007; Tolliver video interview, ca. 1990. Tolliver's hearing was seriously damaged from quinine used for malaria. He was awarded $1.50 a month disability payment in 1945. He later received free hearing aids for both ears and batteries from the VA. He refused to ask for any further benefits until 1996-97 when his wife's pleas led him to ask for full benefits. He then received $2200 a month.

[19] Leatrice Clauson Cooper, "My Brother Swede," an essay she wrote about Vincent "Swede" Clauson's life in 1995 and updated upon his death in 2008.

[20] Fred Hales, undated [ca. 1998] interview with James Fenelon. Casualties of the regiment from Cooper and Smith, p. 328, which notes that this amounted to about 50% of the regiment's authorized size. Many more, of course, suffered from malaria and most had problems from it for the rest of their lives.

[21] Burtell and Walker interviews with Shoptaugh.

[22] Scheer and Tolliver-Pompa communications with Shoptaugh.

[23] Wagner telephone interview with Shoptaugh and early copies of the *164th Infantry News*, (I Co.). Veterans of the 164th were also welcomed at reunions of the 1st Marine Division and over the years several men did attend some of these gatherings.

[24] "Joe E. Brown to Be a Reunion Guest," *164th Infantry News*, October 1963. A nearly complete collection of the *164th Infantry News* is in the collections of the Association Records at the University of North Dakota. The president of the Association is chosen by annual election, and there were 37 presidents in the first 40 years.

[25] It was hardly a coincidence that all three of the men mentioned were journalists whose training inclined them toward the type of "Joe Blow stories" that war correspondents had used during the war to make the war personal to average readers. When all three books, appearing between 1957 and 1961, became best sellers it encouraged publishers to offer attractive contracts for more such works. Studs Terkel's bestselling *The Good War*, published in 1984 in the midst of the fortieth anniversaries of World War II was another such book. Its appearance at a time when the "Depression-War" generation was now retiring simply boosted this form of war memoir.

[26] *The Marines' War: An Account of the Struggle for the Pacific from Both American and Japanese Sources* (New York: W. Sloan, 1948), pp. 87-88. The official Army and Marine Corps history are the previous mentioned books by Hough, et. al, *Pearl Harbor to Guadalcanal*; Miller, *Guadalcanal*; Miller, *Cartwheel*; and Smith, *Triumph in the Philippines*. See also Cronin, *Under the Southern Cross*. These official histories relied heavily on the formal reports submitted by the divisional and regimental commands, which seldom contained information about the ordinary soldiers in the campaigns.

[27] Carl Forsberg, letter to editor, July 18, 1975 (commenting on photo of Clyde Morgan's gravestone published in the June 1975 issue), *164th Infantry News*, December 1975; Jim Winter, letter to editor [1989], *164th Infantry News*, July 1989.

[28] "Please Help Me Get to Know My Father?" *164th Infantry News*, June 1997.

[29] The 164th Infantry Memorial and its dedication are described the *164th Infantry News*, August and December 1994 issues; 1992 flight to Guadalcanal

preserved in video documentary at the North Dakota Guard headquarters, Bismarck and discussed in "Guadalcanal Revisited," *164th Infantry News*, November 1992.

[30] Norman Thompson, letter to Arvid Thompson, December 1992, printed in *164th Infantry News*, October 2000; Edward R. Puhr to the Editor, March 25, 1987, *164th Infantry News*, June 1987; Shoptaugh conversation with Rita Fox, December 2009

[31] Nelson tribute to Baglien reprinted in Nelson's "From Prairie to Palm Trees."

[32] William Tucker interview with Shoptaugh.

[33] Dan Scheer, e-mail to Shoptaugh, March 8, 2007.

SOURCES

The main archival sources for this work are the records of the 164th Infantry Regiment, now part of the 164th Infantry Association Records, Elwyn B. Robinson Department of Special Collections, Chester Fritz Library, University of North Dakota. The original copies of these reports, daily intelligence summaries, maps, and other documents are in the National Archives. Secondary literature and other archival sources used are all listed in the chapter endnotes.

In addition, the author made use of interviews he conducted; and interviews, diaries, letters, and other documents provided by veterans or members of their families. My thanks to those who contributed these interviews, letters and documents:

John Anderson	Clement (Murphy) Fox	Philip Rackliffe
Samuel Baglien	William Freeman	Richard Reid
James Beaton	Elroy Greuel	Walter Rivinius
Emil Blomstrann	Joel Grotte	Ben Rosalez
Melvin Bork	Paul Guzie	Charles Ross
William Boyd	John Hagen	Robert Sanders
Gene Brinkman	William Hagen	Gerald Sanderson.
William Buckingham	Anthony Hannel	Bernard Scheer
Douglas Burtell	Don Hoffman	Russell Schmoker
Walter Byers	Zane Jacobs	Gene Scolavino
Douglas Campbell	William Jordan	Milton Shedd
Melvin Carlen	John Kasberger	James Simpson
Joseph Castagneto	Woodrow Keeble	Rolf Slen
Vincent (Swede) Clauson	Lester Kerbaugh	Eugene Solomon
Crescencio Cruz	Arthur King	John Stannard
Anthony Czarnecki	Edwin Kjelstrom	Richard Stevens
Charles E. Davis	Howard Lauter	Bennie Thornberg
George Dingledy	Robert Love	Arthur Timboe
Eli Dobervich	Fred Maier	Alvin Tollefsrud
Robert Dodd	Albert Martin	Ernest Tolliver
Frank Doe	Marion Minks	Fr. Thomas Tracy
Fred Drew	Donald Monger	William Tucker
William Dunphy	Edward Mulligan	John Tuff
Rudolph Edwardson	Horace Nearhood	Bernard Wagner
Raymond Ellerman	Arthur Nix	Charles Walker
Neal Emery	Ralph Oehlke	William Welander
Philip Engstrom	Robert Olson	Lester Wichmann
Jim Fenelon	Russell Opat	Wendell Wichmann
Dennis Ferk	Claude Parish	Harry Wiens
Warren Fitch	John Paulson	Albert Wiest

Index

Acosta, Joe	362
Aldrich, Les	65, 172, 239
Anda, Stanley	128, 198
Anderson, John	187, 188, 290, 291, 292
Anderson, Vernet	240
Andrick, Rollie	21
Arnold, Gen. William	327, 332, 333, 337, 368, 372
Arrowood, Clair	235
Baglien, Samuel	10, 19, 23, 41, 90-93, 99, 101, 108, 109, 114, 115, 117, 118, 138, 139, 146, 152, 154, 155, 163, 165, 170, 171, 175, 199, 201, 210, 211, 212, 223-225, 237, 399
Baird, Col. LaRoy	38, 39, 46
Baldino, Pete	299, 300
Bandow, Jess	196
Beaton, James	194
Bednarz, Clarence	200
Bennick, Harry	187
Bethke, Lester	322
Bjerke, Bill	143
Blomstrann, Emil	282, 295
Bonderud, Clarence	218, 221
Bork, Melvin	32, 34, 46, 67, 152, 153, 261, 262
Boyd, William	45, 46, 59, 83, 132, 184, 211
Brinkman, Gene	367
Brown, Joe E.	197, 395
Buckingham, William	292
Burns, Bill	185
Burns, Ed	185, 198
Burtell, Douglas	4, 17, 43, 47, 54, 59, 64, 68-70, 76, 83, 92, 117, 152, 192, 222, 238, 245, 273, 274, 278, 279, 284-287, 295, 319, 320, 340, 341, 365, 370, 391, 394
Busching, Leroy	38
Byers, Walter	143, 233, 239
Campbell, Douglas	14, 38, 39, 42, 55, 56, 68, 189
Castagneto, Joseph	242, 243, 256, 310, 341, 376, 388
Christopher, Utah	347

Clark, William	118, 119, 161
Clauson, Vincent	183, 298, 318, 319, 361, 362, 393
Clewes, Thomas	254
Considine, William	60, 241, 305
Cooper, Vicki Ellerman	233, 234
Cortez, Leonard	348
Cruz, Crescencio	348, 397
Czarnecki, Athony	272, 274, 316
Davis, Charles E.	369
Dawson, John	234
Deering, Barney	92
Dingledy, George	258-262, 272, 273, 305, 358, 369, 376, 384, 388
Dobervich, Eli	295
Dockter, Philip	320, 321
Dodd, Robert	37, 76, 83, 152, 238, 309, 370
Doe, Frank	22
Drew, Fred	374, 375, 384
Duis, George	171, 172
Dunn, Floyd	241
Dunphy, William	186, 187
Edwards, Heber	39, 46
Edwardson, Rudolph	262, 284, 286, 297, 370, 385
Eichelberger, Gen. Robert	322
Eisenhower, Gen. Dwight D.	61, 211
Ellerman, Raymond	196, 199, 234
Emery, Neil	131, 158, 191, 221, 240
Engstrom, Philip	77, 83, 92, 97, 156, 174, 175, 385
Ensminger, Walt	196
Fenelon, Jim	36, 91, 153, 156, 173, 196, 203, 384, 394
Ferk, Dennis	34, 42, 61, 77, 118, 182, 183, 192, 219, 291, 292
Flo, Fred	165, 193, 241, 387
Foubert, Kenneth	8-10, 33
Fox, Clement	167, 283
Fox, Lloyd	200
Frederickson, Ozzie	45, 56
Freeman, William	295
Garvin, Crump	241, 255, 271, 273, 274, 298, 305
Gee, Samuel	257, 305
Ghormley, Adm. Robert	3, 86, 87, 89, 99

Goff, Ed	185
Gossett, John	99, 153, 154, 202, 234, 305, 311, 332, 333, 352, 353
Greuel, Elroy	70, 231, 320
Griswold, Gen. Oscar	255, 256, 265, 270, 278, 308
Grotte, Joel	200
Guzie, Paul	315, 316, 320, 321, 338, 339, 368, 373, 374, 385, 386
Hagen, John	13, 20, 22, 44, 45, 54, 57, 62, 77, 83, 103, 108, 132, 133, 137, 167, 168, 184, 185, 238, 239
Hagen, Olaf	44, 45, 56, 57, 238
Hagen, William	54, 219
Hall, Robert	11, 41, 115, 116, 124, 224, 261
Halsey, Adm. William	99, 110, 124, 140, 164, 166, 194, 231, 246, 249
Hannel, Anthony	216
Heller, Owen	125
Hodge, Gen. John R.	240, 241, 270, 273, 278
Hoffman, Don	42, 57, 58, 113, 115, 131
Holzworth, Ray	137
Hyakutake, Harukichi	145, 263, 283
Jacobs, Zane	344-348, 353, 354, 366
Jagears, Park	11
Jeffries, Robert	312
Jordan, William	263
Kachel, Jack	345, 347, 348
Kanda, Masatane	263-265, 267, 269, 272, 274, 279, 281
Kasberger, John	127, 185
Keeble, Woodrow	36, 172, 173, 388
Kerbaugh, Lester	98, 119, 120, 125, 126, 281, 387
King, Adm. Ernest	358, 371
King, Dr. Arthur	81-82, 103
Kjelstrom, Edwin	46, 61
Knott, Ralph	19, 116, 172, 180, 181, 202, 212
Kolden, Chris	291
Krogh, Bill	129
La Fournaise, Joe	212
Lang, Lavern	234
Lauer, Joe	266
Lauter, Howard	81, 83, 137
Lochner, Louis	118, 119
MacArthur, Gen. Douglas	3, 84, 85, 87, 210, 225, 243, 245, 264, 306-309, 311, 322, 359

Mahoney, William	305, 315, 321, 327, 333, 335, 340, 343, 351-353 366, 368, 373
Maier, Fred	83, 365
Manjome, Takeo	326, 327, 332, 335-337
Marseglia, Joe	79
Marshall, Gen. George	52, 53, 62, 63, 65, 66, 81, 85, 210, 211, 234, 237, 358
Martin, Albert	215
Maruyama, Gen. Masao	104-106, 110, 124, 129, 135, 140
McCarten, Larry	365
McCarthy, Kevin	113, 119
McClintock, Harold	198
McConnell, Veon	334, 335, 344, 351, 352, 353
McCulloch, Gen. William	270
McKechnie, Perry	232
McNair, Gen. Leslie	53
Miller, Joe	158, 159, 161, 162
Minks, Marion	316, 317
Mjogdalen, William	46, 305
Montgomery, Christian	181
Moore, Col. Bryant	3, 91, 97, 108, 115, 139, 145, 146, 155-157, 162, 163, 168, 170, 182, 194, 201, 203, 210, 212, 222, 225, 236, 241
Morgan, Rilie	132, 133, 138, 184, 236, 392
Mulligan, Edward	19, 83, 101, 302
Nearhood, Horace	269, 313, 314, 385
Nearhood, Forrest	313
Nelson, Thurston	399
Newgard, George	92, 93
Nimitz, Adm. Chester	3, 85, 89, 99, 139, 223, 246, 306
Nix, Arthur	101, 102, 183, 185, 199, 209, 225
Northridge, Ben	162, 226
Oehlke, Ralph	38, 46, 58, 67, 191
Oie, Satoshi	344
Olson, Sherman	144
Opat, Russell	35, 92, 199, 221
Ordahl, Stafford	36, 241
Osman, Albert	154
Otmar, Joe	114
Overbeck, Harry	156
Panettiere, Andrew	174, 181

Parish, Claude	367
Patch, Gen. Alexander	81, 82, 89, 90, 92, 163, 190, 193-195, 209, 210-212, 220, 221, 223, 237, 293
Paulson, John	11, 18, 33, 54, 61, 67, 71, 73, 102, 118, 137, 139, 160, 161, 202, 242, 279, 280, 298, 299, 341-343, 370, 392
Poe, Lawrence	112, 113, 239
Puller, Lewis Chesty	110-112, 115, 116, 120, 124, 139, 158, 212, 237
Rackliffe, Philip	188
Reeder, Red	234, 235
Reid, Richard	289
Richards, Frank	224-225, 235
Ridings, Gen. Eugene	312, 326, 327
Rivinius, Walter	118, 156, 173-175, 385
Rosalez, Ben	349-350
Ross, Charles	245, 281, 287, 288, 311, 312, 318, 340, 387
Rupertus, Gen. William	145, 151, 157, 158
Sanders, Robert	43, 64
Sanderson., Gerald	54, 70, 71, 217-220, 226, 233, 280, 281, 382
Sarles, Col. Earle	39, 46, 60, 61, 64, 65, 67, 90, 91, 146, 202
Schatz, George	54, 154, 210
Scheer, Bernard	138, 322, 390, 391, 394, 399
Schmoker, Russell	6, 7
Schoeffler, Franklin	189
Scolavino, Gene	289, 290, 341, 375, 388
Sebree, Gen. Edwin	146, 155, 158, 162, 168, 194, 210, 236
Shedd, Milton	296, 305, 311, 319, 358, 385
Shelley, Miles	33, 61, 154, 202, 279-281, 392
Sherar, Pete	257
Shoji, Col. Toshinaro	136, 145, 151, 152, 157
Simmons, Jack	92
Simpson, James	391, 394, 397
Slen, Rolf	247, 306, 371, 372
Slingsby, John	152
Solomon, Eugene	293, 371
Sommers, Francis	190
St. Claire, Gordon	101, 224

Stannard, John	98, 100, 106, 107, 111-113, 117, 119, 125, 126, 130, 134, 137, 139, 158, 162, 190, 239, 387
Starkenberg, Bernard	21, 44
Steckler, Ted	212
Stevens, Richard	71, 72, 76, 77, 126, 127, 133, 134, 168, 170, 171, 184, 197, 254, 311, 387, 388
Sukanaivalu, Sefanaia	288
Swanson, Leo	78
Sweeney, Lloyd	144
Tenborg, Harry	13, 202, 203, 236, 289
Thornberg, Bennie	7, 231, 239
Timboe, Arthur	40, 41, 64, 97, 100, 154, 162, 163, 194, 195, 237, 241
Tollefsrud, Alvin	5-7, 54, 64, 90, 101, 108, 115, 128-130, 136-137, 201, 218, 239, 266, 370, 385
Tolliver, Ernest	392, 394
Tracy, Fr. Thomas	202, 203, 224, 389
Tucker, William	9, 32, 33, 56, 142, 143, 173, 174, 198, 399
Tuff, John	35, 60, 127, 202, 242, 272, 340, 383, 385, 388, 398
Vadnie, Harry	203
Van Tassel, Howard	212
Vandegrift, Gen. Alexander A.	3, 19-21, 87-89, 100, 104, 109, 110, 116, 124, 139-142, 145, 151, 155, 157, 158, 162, 164, 166, 167, 170, 190, 193, 196, 212, 237
Vettel, Carl	150
Wagner, Bernard	31, 32, 61, 67, 71, 198, 212, 218, 220, 286, 297, 388, 395
Walker, Charles	8, 15, 41, 98, 99, 106, 112, 113, 119, 130, 135, 153-156, 161, 162, 202, 233, 238, 241, 242, 257, 266, 273, 278, 279, 281, 285, 286, 296, 309, 312, 313, 332-334, 337, 340, 343, 344, 351, 353, 366, 376, 394
Welander, William	15, 16, 56, 61, 125, 138, 193, 213, 216, 217, 239
Wells, John	266, 282, 295

White, Howard	70, 219
Wichmann, Lester	76, 77, 78, 144, 199
Wichmann, Wendell	60, 69, 70, 335, 388
Wiens, Harry	1, 2, 4-7, 10, 18, 19, 115, 116, 120, 133, 136, 180-182, 192, 197, 212-214, 218, 224-227
Wiest, Albert	9, 11, 16, 37, 38, 41, 42, 58, 114, 115, 120, 126, 127, 174, 181, 212, 221, 222, 242, 297, 340, 388, 399
Wright, Gerald	14, 38, 39, 64, 101, 185, 202
Zumpf, Harold	395